— PRAISE

HUNT FOR THE S

T0273869

'Beautifully crafted, fascinating and unbearably poignant; I totally loved this book.'

Isabella Tree, author of *Wilding*

'For too long, wolves have been the stuff of nightmares. But, as Derek Gow shows with wit and passion, we should be dreaming of their return. Gow's anecdotes will leave you howling – and his historical detective work to track down the shadow of the wolf is as gripping as any thriller. Derek Gow offers not just an elegy for the loss of Britain's wolves but also a clear-sighted case for their reintroduction to these shores.'

Guy Shrubsole, author of *The Lost Rainforests of Britain*

'A wild hunt of a book: savage, romantic yet as unflinching as a wolf's gaze. Gow slyly asks us to consider which are the monsters – wolves or humans?'

Zoe Gilbert, author of *Mischief Acts*

'A fierce, essential interrogation of how nature suffers when we let fear dominate fact. This book bites and snarls with truth yet not without Derek's characteristic warmth and wit. I hope *Hunt for the Shadow Wolf* sparks the change and vision it so deserves.'

Sophie Pavelle, author of *Forget Me Not*

'A dazzling romp through the turbulent history of wolves in Britain. Drawing on years of meticulous research and personal experience, *Hunt for the Shadow Wolf* bristles with trademark wit, passion and pugnacity. We have lived alongside wolves for far longer than we haven't, and Gow reveals their ghosts lurking in our landscapes, our stories and our psyches, and makes it plain that a nation that kills its wolves is one that kills itself. More than anyone, Derek Gow is the man destined to undo that grievous act of self-harm.'

Lee Schofield, author of *Wild Fell*

'This is my kind of hunt. Equipped with a fierce compassion for the natural world and an equally ferocious curiosity about everything he encounters, Gow ranges through field and forest and fairy tale in pursuit of the elusive last wolf in Britain. Along the way he corners some colourful prey: cave skulls, church carvings, folk legends, Bronze Age barrows, wolf-head trees, and the found poetry of old laws and chronicles. I found myself underlining something fascinating on every page.'

Christopher Hadley, author of *The Road*

'With his trademark wit and erudition, Derek Gow masterfully weaves history, biology and myth in telling the epic story of the northern hemisphere's most misunderstood mammal. *Hunt for the Shadow Wolf* is indispensable for any reader, on either side of the pond, interested in the saga of this iconic carnivore – or its future.'

Ben Goldfarb, author of *Crossings* and *Eager*

'*Hunt for the Shadow Wolf* takes our long, complex, heartbreaking relationship with wolves and lays it all out so that we can see exactly what kind of people we were, what we did, and, given time, what we could put right. Using archaeology, DNA analysis, place names, physical evidence in the landscape, old tales, half-forgotten memories and historical accounts, our dealings with wolves are pieced together to present a picture of humanity's abject failure to live with animals that challenges us to the core of our being. Like the man himself, this book is direct, unafraid and at times an uncomfortable read. Not for one sentence, though, is it boring or obvious. So take a deep breath, strap in and be prepared for a rollercoaster of a ride.'

Mary Colwell, author of *Curlew Moon* and *The Gathering Place*

'A wonderful book – a real tour de force.'

Danielle Schreve, professor of quaternary science,
Royal Holloway University of London

'*Hunt for the Shadow Wolf* lays bare the paradox of Britain's relationship with the wolf. Derek Gow shows, as only he can, the history of a persecuted and maligned species that once shared these isles with us. Told with wit and passion, the many stories of "last wolves" and their superhuman hunters, ferocious sheep and child-killers, and prices paid for wolves' heads, are all contrasted with personal knowledge and scholarly research that exposes the lies behind the legends. Wolves are often simply the projection of all our fears: a shade that bears little resemblance to the living, breathing animal.'

Ross Barnett, author of *The Missing Lynx*

'This piercingly beautiful, achingly sad yet fundamentally optimistic book is a triumph. One day we will have wolves back in Britain, and it will be in large part due to the work of this wonderful man.'

Ben Goldsmith, author of *God Is an Octopus*

'Everything you ever wanted to know about wolves and the possibility of their return to these shores, told with passion and persuasiveness.'

Robin Hanbury-Tenison, author of *Taming the Four Horsemen*

'A chewy gem of a book. It holds more than information: it contains knowledge, even wisdom. *Hunt for the Shadow Wolf* is a roving beast all of itself.'

Martin Shaw, author of *Smoke Hole*

'Reading *Hunt for the Shadow Wolf* feels like gathering around a campfire on a chilly night and listening to a ghost story. But this time, the ghosts are just beyond the firelight, listening to their stories being told with accuracy for the first time. Gow has masterfully woven history, biology, social science and legend in a tale about the wolves of Britain.'

Kira Cassidy, research associate, Yellowstone Wolf Project

WOLFHILL
P'MANY WOOLFS
WOLFBURN
BDNE
NCHNA CAVES
DAMPH
LPCH MADDY
WOLF CAMP
SPITTAL
WOLFSTONE
WOLFSTONE
WOLF CLEUGH
WOLF WOLFSTONE
CORNER
WOLF COTTE
SHENVAL
SPITTAL OF GLENSHEE
LOCHAN A'MHADAIDH
WOLFLAND
WOLFGRAIN
WOLFCLYDE
WOLFLEE
WOLF (WOLFHILL)
BRACKLEY
(WOLF FIELD)
BREAGHY
(WOLF FIELD)
WOLF FIELD OF FAIRY HILL
WOLF
HOLES
(WOLF'S HILL)
WODLOW
WOOLTHWAITE
(CLEARING INFESTED BY
WOLVES)
HOWL MOOR
BREAHIG
(WOLF FIELD)
AUGHNABRACK
(HILL OF WOLVES)
CAHERBREAGH
(DEN OF WOLVES)
WOLFHILL
WOLFSTONES WOLFHALL
ULDALE
(WOLF VALLEY)
WOLFPIT
WOLFALL (WOLF TRAP)
UIVETHWAITE (WOLF'S
CLEARING)
WOLFSCOTEDALE
BRITWAY
(WOLF FIELD)
WOLF FOREST
BLEDFA
WOLF
LAKE
WOLFPIT
ISKANAMACTEERA WOLFHILL
(WATER OF WOLVES)
CREFT
BLEEDING
WOLF INN
WOLFELEY HILL
WODIFELL HILL ROAD
(WOLF FEN)
WOOLPITS
WOOFIT (WOLFPIT)
WOLFS
CASTLE
WERN BLEIDDIAU
(WOLF'S) WOLPHY
GLEN
WOLFS FD
(WOLF'S FIELD)
WOOLPIT
WOOFERTON CROFT
(WOLF ENCLOSURE)
SHE WOLF ENCLOSURE
WOOLMANS
WOOD
LITTLEWOLFPITS
(WOLF CLEARING)
WOOLEY
WOEFUL HILL
(WOLF HILL ENCLOSURE)
WOLFHALL
(WOLFFARM)
TREBLETHICK
CARBLAKE
(WOLF PEN)
HOWLEY
WOOLEY HO
(WOLVES' WOOD)
WOOLEY DOWN
(WOLF GLADE)
WOLMER FOREST (WOLVES CLEARING)
(WOLF MERE) WOOLFLY FARM
WOLF CRAG (DEAD WOLVES
IN WELL)
PEVENSEY
CASTLE

D.J. GOW 2023.

HUNT FOR THE SHADOW WOLF

THE LOST HISTORY OF WOLVES IN BRITAIN AND THE MYTHS AND STORIES THAT SURROUND THEM

DEREK GOW

Chelsea Green Publishing
London, UK
White River Junction, Vermont USA

Development Editor: Muna Reyal
Project Manager: Susan Pegg
Copy Editor: Susan Pegg
Proofreader: Jacqui Lewis
Indexer: MFE Editorial Services
Page Layout: Laura Jones-Rivera

Printed in the United States of America.
First printing January 2024.
10 9 8 7 6 5 4 3 2 1   24 25 26 27 28

ISBN 978-1-915294-46-3 (US paperback)
ISBN 978-1-915294-55-5 (US ebook)
ISBN 978-1-645020-44-8 (audiobook)

Chelsea Green Publishing
London, UK
White River Junction, Vermont USA
www.chelseagreen.co.uk

Dedicated to Roy Dennis MBE.

*The most courageous, determined and inspiring
nature conservationist I have ever known.*

— CONTENTS —

A Last Wolf

The Assyrian came down like the wolf on the fold,
And his cohorts were gleaming in purple and gold.

Lord Byron, 'The Destruction of Sennacherib'

L ike a ribbon of silver, the River Findhorn flows
from the massif of the Monadhliath Mountains to
its mouth, wide open and foaming in the spume of
the Moray Firth. As you drive up the A9, before turning
left into the glen of Coignafearn, you pass a sign on your
right for the Cairngorms National Park.

Blink and you will miss it.

Behind it, on a slope standing tall against the sky, is
the stone erected to commemorate the end of a wolf's
life. One of the last of a dying race.

The hunt took place in the short dark days of winter.
Though a ruin, the hero's house still stands. Firm on the
Findhorn facing its fortune.

Dougal MacQueen, who farmed at Pall-a'-Chrocain,
was the most celebrated bard of the Findhorn. Although

he died in 1797, as a child he would have known of the old wolf pits at Moy, no distance from his home. Dug into a steep-sided, natural ridge, which the packs had followed forever, those that remain are overlain with vegetation of all sorts. Ferns and mosses in the higher stone clefts within reach of a deer stretching down. Seedling trees lower still where the light still allows. Stories now lost would have told of their use and of the great wolves and small, which met their ends there. Perhaps as a boy Dougal had listened, crouching shins hunched in the shadows while the peat fires spat and the old men spoke of their victories.

Maybe he knew of the long winding ridge of sand and gravel, dry and deep in the high plateau above the valley's floor. Surrounded by a great bog, this ice-age esker, or 'serpent kame', is prowled still by the gamekeepers who check its centre for a den that the foxes inhabit. But the lair was dug by no fox and the first of that red-russet race to investigate its potential centuries ago must have done so with heart-trembling trepidation. While pursuit above ground by a wolf might have offered a fox a slim prospect of survival, being caught below would have meant a ripping, crushing end. Broken but perhaps with sufficient life left to afford the last litter of wolf whelps born in Britain with a fox-hunting experience they would never live long enough to use.[1] The remoteness of the esker lair made it difficult to destroy and without point in the end, once its last inhabitants had gone.

A Last Wolf

MacQueen was well over six feet in height and, like all heroes of legend, remarkable for his strength, courage and celebrity. He was, as his father had been before him, a stalker who maintained the best 'long dogs' or deer hounds in the country. One day in the winter of 1743, he received a message from his lord, the Laird of Mackintosh, that a large 'black beast', supposed to be a wolf, had appeared in the glen and that it had killed two children the day before who were crossing the hills from Cawdor with their mother. In response to this, a 'tainchel' – a gathering of hunters – had been summoned at a location called Fi-Giuthas. MacQueen was invited to attend with his hounds and, after clarifying where the children had been killed and looking for the wolf's tracks, promised his assistance.

The laird and his retinue had assembled early for the tainchel but by mid-morning were still waiting for Mac-Queen to arrive. At last, through the mist, he appeared like a wraith striding towards them with his hounds trotting at his side. The laird spurred his horse forward to meet him and expressed his disappointment.

'*Ciod e a' chabhag?*' ('What's the hurry?') asked MacQueen.

Mackintosh and his companions chorused impatient and angry replies.

From beneath his plaid, MacQueen withdrew the black wolf's bloody head and tossed it to the ground.

'*Sin e dhuibh!*' ('There it is for you!') he said.

Ecstatic and full of respect for his champion, Mack-intosh bestowed upon MacQueen some land called

Sean-achan so that he would always have means to feed himself and his dogs.[2]

In response to the calls of appreciation from the crowd, MacQueen casually recounted his tale:

> As I came through the slochk [ravine] by east the hill there, I foregathered wi' the beast. My long dog there turned him. I buckled wi' him, and dirkit him, and syne whuttled his craig [cut his throat], and brought awa' his countenance, for fear he might come alive again; for they are very precarious creatures.[3]

It's a great story, and while the naturalist David Stephen felt that it 'ought to be true, and may well be', MacQueen's wolf has been disparaged by others.[4] Jim Crumley, in his book *The Last Wolf*, considered that, 'It was too late in time ... they had been done to death a century or more before. There were not and never have been any black wolves in Britain.' Others have gone as far as to call it a lie, but this, I think, is very wrong.[5] As we shall see, there are sober accounts of surviving wolves right up until the early 1700s and plausible tales long after 1743. Although we will never know what hue his wolf was for sure, there may also be an entirely reasonable explanation for its dark colouration.

I believe now that wolves lingered on in mainland Britain well into the eighteenth century and that in Ireland they survived for perhaps a little longer.

And I believe the account of MacQueen's bold fight.

A Last Wolf

The Gaelic tradition of oral tale-telling in the Highlands was once very strong. Although weakened, like Irish Gaelic and Welsh, by the onslaught of English, a story of this sort of consequence is unlikely not to have had some basis of truth. While the foregoing in slightly varying forms was the only version I had ever seen until I began to assemble my material for this book, it transpires that corroboration exists. It comes from Duthil parish near Elgin and is contained in *The New Statistical Account of Scotland, Volume XIII*, published in 1845. While it does not specifically name MacQueen, it states that the wolf killer lived in the eastern part of the parish of Moy, which is where his home was.

Its background detail is absorbing.

In the early 1700s, the forest of Duthil was destroyed by a great forest fire. Only a single wolf from the several that escaped survived the hunters who pursued them. The account tells of how an 'overgrown animal' fled to Moy before making its presence known:

The inhabitants had a fearful warning of its being among them, by its killing a woman and her infant child. As soon as the laird of Mackintosh heard of this melancholy event, he summoned his vassals … Their intentions were, however, anticipated by a daring fellow, that lived in the eastern extremity of the parish, who, as he was on his way to join the rest of his clansmen, was met by the very animal in question, in an exceedingly narrow path in the face of a rock, called Creig a chrochdan … The man, by a well-directed stroke of his club, brought his foe to the ground … Thus perished the last of the native inhabitants of the forest of Glenchearnich.[6]

Names change with ease and the modern map name of Creag a' Chrocain, which means 'craig of the crook', is simply a reflection of this. The hill, with its narrow footpath, has moved not at all from its foundations and remains where it always was above MacQueen's old farm. Between Fi-Giuthas, which means 'bog stream of the fir tree', where the clansmen met, and MacQueen's home at Pall-a'-Chrocain, is a place called Caochan a' Ghubhais, which translates as 'wee moor stream almost hidden by heather or the undergrowth of the fir'.

Although Gaelic is a language of poetic flamboyance, Caochan a' Ghubhais and Fi-Giuthas in essence differ not much in meaning at all.

Was Caochan the point where Dougal's battle occurred?

Whatever others think, the modern MacQueens still believe Dougal's story and retain on their clan crest three dark wolf heads with long crimson tongues.

Factual or fictional character, his wolf will never die and he remains their hero forever.

Some years ago, on an unusually idle afternoon, Roy Dennis, the great restorer of raptorial birds to Britain, was watching an international rugby match in the comfort of his front room in Abernethy. One commentator shouted, 'Bloody well done, dogcatcher,' as the Australian captain scored a try to wild applause. His assistant on air asked, 'Why did you call him dogcatcher?' And he replied, 'Because he's a MacQueen, mate, and his ancestor killed the last wolf in Scotland.'

Hunting the Wolf

> I walk
> with that wolf
> that is no more.
>
> Toshio Mihashi, Japanese haiku

To commence a hunt for a species that has passed may seem pointless. If you're a hunter, you need a quarry and, if your object is now nothing more than myth, then what reason can there be for its pursuit?

No rack of horns on the wall. No umbrella foot basket. No rug to fade in the study with the cracked ivories of its incisors glinting in strobes of dust-filled sunlight at midday.

It has never been my ambition to seek trophies. While stuffed creatures, shells, bones, nests and skins of all sorts adorn the walls of the classroom on my farm, I have acquired them quite casually when the opportunity arose. The single wolf skull that resides there came from a male American timber wolf, which died very suddenly in a Scottish zoo. One day it was hearty, the next septi-caemia from an ulcerated tooth took its life as it curled up under a spruce tree to sleep. I found it immobile, bedecked with diamanté spider webs in the dew of the following day.

I have kept and handled tame wolves of the small brick-red European form and the much larger, grizzled North American timber sort when I worked in the past in zoological collections. I know that when you bury your face deep in their tawny or steel-blue fur they smell of bone marrow, and that when playful juveniles roll over and you get your lips in close enough to blow raspberries on their quivering pink tummies, they quite readily wee themselves with joy.

I helped to hand-rear two females – Nadia and Mishka – in a wildlife centre where I worked in the early 2000s. They were European wolves whose parents had been rescued from an Eastern European fur farm and lived in my house in Kent for a while. Though always endear-ing, they were incredibly destructive. Ripping, pulling, tugging was their greatest of pleasures, with clothes and furniture alike. Novelties such as toothbrushes or keys unwarily placed would be swiftly ingested. Once inside a wolf, only one option remained for any item of value.

To wait for around eight hours with a clothes peg and gloves to search through whatever came out.

In the evenings, they liked to lurk in the shadows of the upper stair landing so that when you left the comfort of the sofa and fire in the front room after an early-evening snooze to make your way yawning to the kitchen for a bedtime cup of tea, they would launch their attack. There was no threat in their leaping ambush, just a wolfish sort of rapid licking and love as you picked yourself out of the carpet again.

At that time, I had three male collies with whom they would willingly play. When they were small, the merest of growls from the two stronger big dogs when it all became too much would reduce the wolves in an instant to supplicant fawning submission. Sprocket, who was grey and white, was, however, a different proposition. Although a sweet soul, his birth had been slow and his mental development was, without doubt, affected by this. The wolves knew this well and, as they got larger, would single him out, force him into a corner and then stand over him looking down. As it became very obvious that one day soon Sprocket was going to become a light wolf snack, they were banished to the garden of my next-door pal, Sarah.

I became interested in the history of the wolf in Britain in a casual sort of way. I was working with many other species at the time and, although it never really struck me then, all I ever found in superficial accounts of their end were scattered stories of a repetitive sort. Great human heroes grappling in lone battles against dark and deadly lupine foes. Tales without credibility told perhaps to haunt or entertain but long past any point of worth. None of their content, given the simple understanding of the wolves I knew, made much sense. Unless goaded well beyond any limit of tolerable endurance, wolves were shy, supine, curious, sneaky at their most malicious, and just boisterous if they liked you a lot and you liked them.

Nothing much in contemporary literature contradicts this simple understanding, but could we really over time have handmade a foe of their sort without any basis in fact?

———

When my journey began to reintroduce the beaver to Britain and I undertook many European trips to view the rich waterworlds they create, wolves were a nothing. They did not exist in most of the landscapes I visited a quarter of a century ago. I recall a field trip to the Netherlands with an agricultural college in Hampshire to look at an engineering project near Nijmegen. As we viewed the steel-grey konik ponies and black Galloway cattle contentedly grazing, I asked the project manager accompanying us if, as part of his ambitious rewilding agenda, he would ever consider reintroducing predators. I recall he laughed before saying no and stating it was his belief that, as more green bridges covered with trees linked to river corridors were built or other landscapes were recovered for nature, in time wolves would arrive of their own volition.

As the nearest wolf population was nearly a thousand kilometres away in the military ranges of Saxony, I dismissed this assumption as fanciful. When several years later a young wolf was killed on a motorway equidistant from the Dutch border and its east German birthplace, it still seemed far-fetched that, long after the last was killed in 1869, they could recolonise the Netherlands. Now, of course, they have and, despite several road deaths and a number of illegal killings, in 2023 there are believed to be nine packs – wolf pairs that have borne cubs – seven of which, plus a scattering of single wolves, are living without great issue in the sprawling pine forests of the central Veluwe.

As the wolf has returned to France and Germany, to Belgium and the Netherlands, to Luxembourg, Lichtenstein and Switzerland, more old beavering chums have told tales of their comeback. Some sad about their strange sudden deaths or disappearances. Odd incidents with tame kangaroos. Of weeping schoolteachers and the hatred of wolves associated with the worst politics of the rising far right.

But there are governments who have banned their killing and many people of all sorts who have decided to offer them welcome. Wolves are not easy to live with. Although wolf-proof fences can be erected to protect sheep in small fields, for large roaming mountain flocks guard dogs and corrals not used since the Middle Ages need to be re-employed and rebuilt. Foals and calves can and will be attacked as wolf numbers increase and, while there remains little evidence that they pose any real threat to us, the fear of the wolf that many individuals carry in the darkest portals of their hearts remains all too alive.

Perhaps we resent them for being a life force that's ungovernable by us.

The English Channel, deep and wide, ensures our sanctity for a wolf-free future here in Britain. If we ever wish, like the Coloradans have just done, to acquire them once more, they will have to be captured and crated from European forests. While, of course, the draconian unions of the farming elite will say no and persuade their pals in power perhaps to back them for a while, there are others in increasing number who long for a different future.

More rewilding estates and progressive farm owners are relaxed. Foresters, who wish to bring deer numbers down and so reinstate a natural equilibrium between landscape and wildlife, know that the materials used for the fences they erect now to deny deer access will only do so for around fifteen years and that, one day, an alternative must come. Other voices in wider society are also rising. Large conservation groups, journalists and writers, sober scientists, young filmmakers, poets and children. In short, people of all sorts are asking: why not?

Many have come to realise that the story of the wolf's eradication from Britain was simply a curtain-raiser for the sheep and the deer, which, in ever-rising numbers, have flayed our uplands bare.

One day, perhaps the slim candle of hope that's smoking slowly may flare fiercely into a torch of ambition.

My quest to find out what happened to the wolves that were once here has become both a mission and delight. As one story faltered or came to a conclusion, another would beckon from the page of an elderly account, a hint from a place name or through the chance encounter with an individual who knew more. Several times, a sentence that I started to explain to a listener would be finished by them with information of which I knew nothing.

For what it's worth, the International Union for Conservation of Nature, in its official reintroduction guidelines, requires that both the cause of a species' extirpation and the absence of the initial drivers of its extinction are understood before any programme of restoration begins. Although we know that we killed wolves in Britain because they ate our sheep, that is a simplistic understanding without any depth of complication.

A better account of what happened in Britain to the wolf might therefore constitute a beginning of sorts for any movement that sought their return.

So, shall we commence?

———

This is the story of the wolves that once lived in Scotland, England and Wales by way of Ireland. It's about what we knew of them before all that remains turns to dust.

I believe now that those of us currently living in Britain did not miss the wolf's time by much. Perhaps as a fragile presence they endured until the early 1800s in Ireland at least, although other accounts follow later from the mainland. If you consider these to be credible, then a last wolf walked this land in a time span of not long ago.

During every warm gap in the ages of Ice that passed in succession for five hundred thousand years, wolves remained if they could in Britain or trotted back when the glaciers and snows flowed away. Around eight thousand years ago, that all stopped when a gigantic landslide in Norway triggered perhaps the biggest tsunami ever recorded on earth. The waves that struck what was north-east England drove twenty-five miles inland, and when their wrath abated, what had been a low-lying plain became the bed of the North Sea. In the south, the swamplands, which could be traversed, flushed away to become the English Channel.

For the people who survived, it is possible that wolves had some worth. Wolf kills could be readily found and scavenged if you followed the cronking bands of black ravens gathering for a feast of their own. Tamed cubs might grow into staunch hunting companions and many cultures worshipped them for their strength, stamina and destructive ability. The Celts believed that they were descendants of the great Roman god of the infernal regions, Dis Pater, who swathed himself in wolf skins, while his companion, the horned Celtic god Cernunnos, walked with a wolf on one side and a deer on the other. Their chiefs sat on wolf skins when eating their meals and laid them as coverings on their floors. Some wore wolf heads on their helmets as a sign of individual courage or interbred captive wolves with their war dogs to create terrible assailants.[1]

While we will never know if the first farmers who arrived two thousand years later were respectful of wolves in the way modern Deccani shepherds are for the service they afford by removing the sick sheep from their flocks, we do know that our ancestors erected defences.[2] Tall walls of stone on the highlands of Cornwall over-looking the western seas enclosed many acres of good grazing and water while simpler stone corrals called cottes in remoter landscapes offered both the sheep and their shepherds safe refuge at night. Crannogs in the Lakelands with connecting underwater causeways and high wooden fence pales set in raised banks around medieval settlements like Kibworth in Leicestershire all had the same purpose.[3]

To keep the wolf at bay.

The earliest illustration of a wolf hunt in Britain dates from an eighth-century panel of the St Andrews Sarcopha-gus and depicts a shield-carrying huntsman on foot, armed with a spear and assisted by a hound attacking a giant of a wolf.[4] Although there are no references to wolf hunting in extant Anglo-Saxon documents, it must have occurred but when William the Bastard killed Harold on the battlefield of Hastings in 1066 his triumph brought a new world order. Hunting was elevated to the most precious of noble recreations and wolves as adversaries were, with wild boar and deer, protected by the royal guarantee of the Forest Laws. Designed by the Normans to 'leave the English nothing but their eyes to weep with', no leaf litter or wild fruit could be gathered.[5] No firewood could be gleaned nor any other product taken from their forest. Individuals who merely disturbed a deer might discover their weeping days were also over when their eyes were declared forfeit.

When a bored Victorian clerk in a London office was tasked with translating a Latin text regarding ancient land rights from a period dating back to between 1272 and 1307, he would hardly have tingled in anticipation at his task. As sentence after monotonous sentence unfurled, a surprise lay in store. Suddenly, with a start, although he understood the meaning of the word, he questioned its use in pencilled exclamation: 'free chace for hares, wolves (?) and wildcats.'[6]

Could it really have been that wolves once roamed the tame lands of Dorset?

While the Romans may have been the first to export the white wool of their hornless sheep from Britain, by the mid 1400s it had become the most important of national assets. Wool paid for churches, castles, wars and treasure chests. Though its value was to peak by the century's end, its trade as either a raw material or broadcloth was to remain of significance for the next three hundred years.

The early medieval churchmen, who were sheep keepers of note, sought to maintain their bleating hordes with as little issue as possible. They had no wish to tolerate the depredations of wild animals and ensured through preaching or persuasion that every hand that could be was turned against them. Minor criminals had their sentences remitted if they brought the authorities severed wolf heads or tongues, while listening congregations believed resolutely in their unquestionable evil as the scatterers of the 'good shepherd's' flocks.

Once society had been taught to hate the penchant of wolves for robbing graves, their willingness to clear up the battlefield dead and to excavate with ease the possibilities of a plague pit's depths damned them still further. Beyond the rim of a camp fire's vision, their chorused howls and demonic crescendos haunted the night and drove those they taunted to seek out their lairs by day. After stabbing, beating, flaying, burning and poisoning wolves for centuries, we destroyed their age-old sanctuaries, dismissed them in fables and told tales of our courage and their cowardice.

It's how victors forever frame those they defeat.

———

There are many 'last wolf' stories in Scotland. Two miles north of Brora in Sutherland, laid into the slope of a bank is a stone that marks the place where a last wolf in Sutherland was killed by a hunter called Polson in or around 1700.

Its background is violent.

Polson, an old hunter who was well experienced in the destruction of wolves and accompanied by his son and a herd boy, found what they thought might be a wolf den with a narrow entrance in the valley of the Sledale burn. As Polson could not pass down into where the wolf might be, his son and the boy descended into its depths and found a litter of cubs in a chamber strewn with the bones of their prey. When on Polson's instruction they began to stab them with daggers, the mother wolf, who had been hiding outside, tried to slip back into her den to protect her offspring. As she did so, Polson grabbed her by the tail and, when his son asked why the daylight from the tunnel above had been obscured, Polson replied, 'If the root of the tail breaks, you will soon know,' before killing her with his knife.[7]

I have my own wolf story.

When I was small my grandmother would always insist on map reading for my long-suffering mother on our Sunday afternoon excursions. As she got older and her eyesight faded, she became ever more insistent that her directions were right and that circumstances outside the car – trees that moved, a cow that looked at her in a funny way or a child that farted in a pram – were to blame for the confusion she caused. Her moment supreme occurred on a drippingly hot day trip to Dundee to visit relatives in my mum's tiny tin can of a black Austin car. After passing the same set of road signs for the fifth time, my sorely vexed parent pulled over at the roadside and stopped to discover that Gran had been following a marmalade stain on the map. She was sacked as navigator and, as my brother and I fought in the back seat, in desperation to entertain us she told her tale of the wolf.

Gran said it was the last of its kind.

One day in the time of long, long ago, a woman with her children walking towards the small town of Biggar, in the flat lands of the Fleming lords where we lived, were attacked by a wolf. Perhaps the children were killed, perhaps they were not; in any case, the doughty response of their mother was to whip out a pancake griddle and bash her assailant over its hairy head.

Now, a Scots pancake griddle is either a flat iron disc with a simple folding handle, also of iron, or a large square cast-iron flat pan with a handle attached. As we shall see time and again, this deadly piece of weaponry was a critical accoutrement for travelling Scots 'wimmin' in wolfen times. Although I have no clear evidence of either design's utility as a means of execution, I think the latter might be best and, assuming this fiction, her victim died promptly.

People were happy and the place where it met its end was called Wolf Clyde.

It's a story that many of my schooltime chums recall being told by their parents in turn. No more detail that I recall was ever added.

But there's a problem with my grandmother's account.

Was her wolf fact, fiction, or a melange of both?

In 1431, the place where Wolf Clyde is now was written 'Wolchclide'. In a map from 1583–1596 it's marked in a manuscript as 'Outhclyd', which has also been explained elsewhere as Wathclyde.[8] 'Wath' translates either as a place with a river ford or a place with a fort. It wasn't marked on any map as 'Wolf Clyde' until around 1773 and it's not until 1816 that 'Wolf Clyde' was recorded.[9] Only in 1894 did one D.P. Menzies record that: 'Tradition says the name of Wolfclyde arose from the last wolf slain on the Clyde having been run down there and killed by a Menzies.'[10]

Although I can't recall now the appearance of the older brother, there were at least two Menzies in the primary school I attended in the little Scottish village of Broughton in Peeblesshire. My pal Stuart was soft and chubby. Round, tender and soaped smooth by his mum. Not much of a martial sort, had he encountered a wolf at the age of nine it's much more likely that his role would have been that of a toothsome morsel than a tormentor, unless of course his bones lodged fast on their way down its throat.

The seed of interest sown by my grandmother set deep roots and, over the years, as I travelled to wolf lands in both North America and Eurasia, and began my zoo career, I began unconsciously to pay more attention to wolves. When time allowed, I dug deeper for further information.

The best account to date of the history of the wolf in Britain was written by James Edmund Harting in 1880. His *British Animals Extinct within Historic Times: With Some Account of British Wild White Cattle* has in more modern reprints a nice grey-tinted lithograph of a wolf gnawing a pile of what could well be human bones on its cover. In 1974, an author called Anthony Austen Dent added further detail in his work *Lost Beasts of Britain.*

Both are worth having if you have an interest in nature and Dent's volume, slim, small and black with a gold wolf woodcut on its front, can still be unearthed in second-hand shops. If you see, it buy it. It is an acquisition you will never regret.

These books provided the physical locations of the wolf traps that I then made time to visit and the place names that recalled the wolves once being where I sought for any fragrance of them that remained. I travelled to see the church tombs of their killers and the memorial stones erected where a wolf once bled out the last of its life. Beyond excursions of this sort were more dusty books. Old clan accounts containing the remains of readers who had died in their depths from tedium. Letters from the archives of *The Field* magazine, which afforded a few snippets of surprise. Scientific or archaeological studies with indisputable facts and newspaper stories of varying sorts.

Vermin records, maps, notes in estate ledgers or translations from works in Old Welsh or Gaelic all followed on.

I asked many folk for their wolf stories. Most had nothing to say but laughed, quite a bit. Others slunk off to relate of my pending insanity, but just a few, after pausing, said something that was worthy of note.

Roy Dennis helped with the detail of the fine fight on the Findhorn, while the sage that is Sir John Lister-Kaye snuggled round a whisky to espouse a more amusing rendition of yet another wolf being despatched by a pancake griddle.

Lots of friends helped by providing romantic fables and directions to the work of female Edwardian poets who had visibly suffered from too much parlour time on their hands.

Several appeals on Twitter produced surprising leads.

As my absorption with the wolf grew, it began to affect my behaviour. I now compulsively pick up books of all kinds on social history, costume, folklore, witchcraft or long-past plagues. If under W at the back there is a mention of wolf, I start to twitch, buy it and return to consume its contents in the depths of my garden office. Though most yield little of worth, I have learnt a lot about the design of medieval gussets and how the implantation of he-goat testicles in the scrotum of impotent men in historic times failed quite utterly to address their condition.

Wolves are just wolves. Nothing more or less. Ours, like all those that remain to roam other lands today, were neither ghouls nor fiends, but predators of power and guile that opposed the way we wished to use the land and vexed us quite grievously by not dying out when we felt they should.

In the end we took everything they had.

Their life-lands and prey.

Their teeth, blood and bone.

Their strong hearts.

I would like to tell you the story of the wolves that were once and of what they might be again if we can ever wrench ourselves free from the past. Gather round now if you will. Close your eyes and hunch in the warmth beneath the peat smoke. Let the fire's golden glow still your mind. Though the tale of this wolf hunt is torturous, if the wind blows outside and a howl can be heard, rest assured it's a ghost of no substance.

CHAPTER ONE

The Iron Wolf

They left behind them to enjoy the corpses,
the dark coated one, the black raven,
the horny beaked one, and the dun coated one,
the eagle, white from behind, to enjoy the carrion,
the greedy bird of war, and the grey animal,
the wolf in the wood.

'Battle of Brunanburh', Anglo-Saxon poem

I f you speak to archaeologists like Professor Greger Larson at Oxford University's School of Archaeology, not much physical evidence for the existence of 'modern' wolves after the last ice age remains. Although an assemblage of gnawed and fractured deer bones from Rawthey Cave in Cumbria may have been left there by wolves in the thirteenth century, no clues of any other sort exist to identify their gatherer's identity. While wolf bones from the Middle Ages have been identified in excavations at Cromarty Castle in Nairn, most of the other physical evidence for British wolves is either of unknown provenance or of considerable age.

None of the sharp, fanged skulls discovered in caverns, unearthed from peat bogs, pits or crypts in recent times have been identified as belonging to anything other than large dogs and the many wolf relicts recorded quite widely by Victorian antiquarians have all, it would seem, been lost. When the railway tracks from Hereford to Brecon were installed around 1862, there were rumours that the skulls of some hundreds of wolves were found in the breach of a gap near Clifford Castle to the north of Hay-on-Wye.[1] In Bury St Edmunds, where the western tower of the abbey once stood and an estate agent's premises now resides, a large pit containing many skulls was likewise supposed to have been discovered. More skulls found in a well at Pevensey castle in Sussex in the mid 1800s or during the construction of the Walthamstow reservoirs in the 1860s cannot likewise be found now.

Any paucity of physical proof, however, cannot always be taken to infer absence. No fossil bones of wolves, for example, have ever been found in the Netherlands, although they were known to exist there well into modern times. For some species, bone rarity means that when fresh material does come to light it can implode firmly held dogma. For example, a single accidental find of mammoth bones in 2009 near Condover in Shropshire moved the presence of elephants in Britain closer to the present day by seven thousand years.[2] If we can't exactly define the timescale for an elephant, how on earth can we do so with accuracy for anything smaller?

But it's hard to believe when you consider all the other sorts of evidence that wolves died out near completely when the last legions of Rome sailed east. Illustrations of them tumble in later centuries from rock carvings, statuettes and lavishly illuminated medieval books. A few of the great traps to catch them remain. On windswept moors 'wolf stones' endure as a memory of a hunt long forgotten or a grey enemy slain. If no wolves survived well beyond the Middle Ages, then why the records of bounty payments for their capture or death? Why the customs records from the sale of their hides? Why the established hunting seasons? Why the keeping of savage hounds to pursue them? Why the poisoners' potions?

The wolf as a real beast in Britain has slunk so far from us now that much of the written material that describes them is limited and brief. Although the children's game 'woof and lambs', where a school flock of chanting kids were sought out by a 'shepherd' who led them to safety while 'wolves' circled round, was still played at Ingleby Greenhow in North Yorkshire until 1914, memories of this sort are now nearly unique.[3] In the mountains of modern Romania, many sheep-farming families still relate by a night fire their stories of wolves. While once there must have been similar here with tales told of or by their notable killers, like the wolf-slayer Jack of Badsaddle who died in 1375 at Orlingbury, all details of their exploits are now beyond recollection.

Lost forever in a forest of forget.

———

Sometimes our soils cast up wolves we have buried.

The statue of a muscular wolf from Woodeaton village in Oxfordshire consuming a human victim head first. Kept in the British Museum, it's believed to be a Celtic idol of worship, which, although darkened by its centuries in the earth, is made from a green copper alloy that once glimmered gold. Nearby in another glass case is the ornate bronze scabbard of a gladius from the first century CE depicting Romulus and Remus, the twin founders of Rome, being suckled by their she-wolf mother 'Lupa', which was discovered in the sediments of the River Thames near Fulham.

In the Wiltshire Museum is a Bronze Age necklace made of wolf and dog incisors cross-cut, sectioned and polished into delicate, mother-of-pearl-like slivers, which was excavated from a Bronze Age bowl barrow near to the village of South Newton. Tiny holes bored through the blunt roots of the teeth to enable them to hang from a cord would have left their pointed tips dangling down in admirable ensemble. In an age of great intricacy in jewellery design, a work of this sort would have been both precious and eye-catchingly desirable.

The 'Ardross Wolf' was carved by an artist of the painted people in the sixth century CE. On display in a glass case in Inverness Museum, in simple lines with whorled leg tops, ears laid back and tongue down, it flows through buff sandstone in exquisite relief. As a well-observed work there can be no doubt that its creator understood his subject's shape. A muscled front leg shape curving tight into the dew claw above its foot, a drooping tail and downward-sloping hips all depict a magnificent beast in the prime of its life.

Wolves, deer and snakes are all engraved on the Shandwick Stone between the Dornoch and Cromarty estuaries, while another carving at Darnaway Castle near Moray depicts a wolf hunter pulling a bow on his quarry. Nearly eight hundred years after their creation, Hector Boece (1465–1536), who was the first principal of Aberdeen University, attributed these ancient memorials to a time when 'engraved images of dragons, wolves and other beasts ... put the deeds of noble men in memory'.[4]

The Venerable Bede (ca. 673–735) was one of the greatest Anglo-Saxon scholars. Although he lived and died between the twin monasteries of Wearmouth and Jarrow in the North East of England, he was one of the first to write about wolves. In a description of 'Anderida' or Ashdown Forest in faraway Sussex, he observed its landscape to be 'all but inaccessible and the resort of large herds of deer and of wolves'.[5]

Around three hundred years later, another cleric, William of Malmesbury (ca. 1095–ca. 1143), spoke of the first organised effort by a king to free England of their presence. In 937 CE, King Athelstan defeated an invasion force of the Scots, Welsh and Danes at the Battle of Brunanburh. Although the site of the conflict is now unclear, what is known is that the defeated Welsh kings agreed to pay a handsome annual tribute in recompense to the English. When his nephew Edgar became king in 959 CE, it was he who instructed that Athelstan's initial choice of gold and silver should be remitted for wolf skins instead.

William recorded that:

The happie and fortunate want of these beasts in England is universallie ascribed to the politike government of King Edgar, who to the interest of the whole countrie might once be cleansed and clearelie rid of them, charged the conquered ... to paie him a yearlie tribute of woolfes skinnes, to be gathered within the land ... none of these noisome creatures were left to be heard of within Wales and England.[6]

While Edgar's attempt to rid England of wolves failed, his effort to do so is acknowledged to this day on the west face of Lichfield Cathedral where his statue bestrides a wolf's head.

His tribute was not a commercial transaction but a penance. A totem to the Welsh of their subjugation by a hated overlord. The autumn would have been a good time to kill when there were adults with a season's crop of cubs. Though physically well-formed, these would have been naïve and clumsy. Without the guile of their parents, they would have been easy hunting. Compliant and submissive when cornered alive. Offering little resistance to the hounds if a running end brought them down. Their bodies in tally would have made payment easy.

The roads may still have been passable before the rains turned all into turmoil. Perhaps horses in pannier trains carried them in. Maybe ox carts piled high. Skins would have been easy, tails or the ears required by the Black Prince four hundred years later easier still. Whole carcases pickled or skulls left to dry would have required more space. More time and trouble. More expense. More bother.

These factors would have troubled the receivers not at all.

Theirs was an exercise in inflicted misery.

Was it a wolf or only a dog? Hands on daggers when arguments rose.

Wolf's teeth as talismans.

When the counting was done and agreed, the English brought their wolves home. While items of worth as winter robes, bed furs, cuffs or medicinals may have been utilised, the diaspora was taken to Woolpit in Cambridgeshire and burnt.

———

More than the legacy of any other native creature, like the fallen petals from a wild fruit tree, wolf place names bless the sediments of our land. Woven into its warp, their power to invoke imagery is strong. On the west coast of Scotland, *'Lochan a' Mhadaidh Riabhaich'* (the wee loch of the brindled wolf) lies like a sapphire twinkling in the green summer glory of Ardnamurchan. Do its waters recall still the memory of a dappled wolf lapping its depths ice cold at the end of a long summer's day?

Had it paused to look down and consider its reflection, as animals often do, it could never have known that the wolf looking back would one day be beyond possibility.

A good pal, Nick Mott, who works for the Staffordshire Wildlife Trust, provided a copy of an Ordnance Survey map from 1878/79 of Wolfscote Dale in Staffordshire, which showed the many sheep folds or cottes that gave the land its modern name. With a single exception, which spans the front of a cave, most of the structures outlined on Nick's map had only three sides: a short front section containing an entrance, while the other two longer walls ran down to the River Dove. Although wolves can swim well – a population that once inhabited Wolf Island on Lough Gill in County Sligo must have swum there to do so – an aquatic attack would have been less likely, and this design allowed the sheep to water at night as the watchers guarded at ease. While this kept the sheep safe, the restriction it imposed meant a loss of night grazing and less milk for cheese.

Bothersome wolves.

At Wolfpits Farm in Radnorshire, the old famer Roy Jones, who had lived there all his life, had always thought his property's title to be wool- rather than wolf-related.

Familiarity with a lifetime with sheep and a complete incomprehension of wolves make his understanding easy to follow. But it's a widespread misinterpretation. If you search for the older origins of modern names, Wooldale valley in the West Riding of Yorkshire becomes wolf dale, Wool Pit in Suffolk, wolf pit.[7] While the names of Wolborough, Wooladon and Woolwell are all likewise masked by the produce of the flock-masters, well over two hundred clear place names such as Wolfhopelee, Wolfhills, Wolfhole, Wolfcleugh, Wolf River, Wolf Ford, Wolf Kielder, Wolf Crag, Wolf Pit and Wolfcotes remain obvious.[8]

Although other references to their former presence are less transparent, when you understand the intertwining of their acoustics – Howl Moor, Howl Common, Howley – with past cubbing dens – Whelpdale, Whelphill, Whelpo, Whelprigg, Whelpside – and their further corruption with age – Wooddale, Wolmer, High Woofhowe, Woofel Hill – the origins of many more are unmasked.[9] In the older languages of Britain they abound. Bleidire in Gaelic translates literally as a 'mouth or mouther' in description of the shearing power of their jaws, while Call-Glaoidheaman was used to describe their double barks or cries.[10]

Set among the Arrocher Alps of Argyll, Ben Vuirich rears near one thousand metres up from the banks of Loch Lomond. Although most linguists agree that its old Gaelic name of Beinn Bhuirich means 'hill of the roaring', this was believed in modern times to refer to the autumn rut of the red deer stags. While the deer remain, their bellows are only a fragment of a once more complex cacophony. Earlier poets recorded that Bhuirich's roars were howls in soaring chorus.

I see Ben Ghlo of the pointed tops,
Ben Bheag and Argiod Bheann,
Ben Bhuirich of the great wolves,
And the Brook of the Bird's Nest by its side.
Old Deeside ways[11]

Roy Jones said there was nothing now to see on his land. No circles. No depression. No odd features at all. The modern farm is all that remains now of a cluster of smaller steadings that once comprised the three different households of Little and Upper Wolfpits. Although their ruins were visible in the 1970s, they now lie entombed under the concrete of modern farm buildings. No record beyond his memory marks their being on a map.

After talking for longer, a surprising thought.

As he warmed to the idea of a wolf explanation, Roy recalled reading many years ago in the nearby church at Clova of a legend that a 'last wolf' was pursued from the forest of Radnor and killed in the time of the Tudors in the pit on his farm. Wolfpits Farm is about ten miles to the south of a site on the other side of Radnor Forest, called Bleddfa. Blaidd, pronounced blithe, is the Welsh word for wolf and Bleddfa translated into English means 'abode of the wolves'.

Although Roy could not visualise how a wolf might be driven to a specific location, it's a plausible prospect. The erection of temporary barriers called hayes and made of wood, wattle, cloth draped over staves or fladry – small flags on ropes through which wolves will not willingly pass – was a common medieval technique widely used to drive massed herds of game, most commonly deer, towards hunters waiting in ambush. As these structures were created from degradable materials, nothing of them now remains, although we know that one described near Donnelie in Warwickshire was half a mile in length and a similar distance in breadth.[12] In Herefordshire, the old place name of Wolphy and in Cheshire, Welfehay, both imply that a wolf 'haye' or fence system of some form was once used to direct wolves to their deaths.[13]

Sometimes once you have found one story, another follows in later support. Roy's farm is only seven kilometres from the source of the River Edw where, in an early parish register from around 1910, the Reverend D. Edmondes-Owen recorded that 'Late in the Tudor period, the last wolf was killed on one of the steep slopes above the river Edw.'[14]

Might this wolf and Roy's be the same?

———

Russell Coope was a giant of a small man. With a flaring beard set in a form best promoted several centuries ago by Sir Francis Drake, he was internationally famous for his work on reconstructing climate change from the remains of ancient beetles. While very many types of beetles over time have changed very little, their distribution has altered hugely according to climate fluctuation. When Russell made a chance discovery of what looked like modern beetles in ancient sediments in the Chelford sand quarry in Cheshire, he was told by a senior colleague, 'Don't be silly, Russell – they are modern ones that have just crawled in there to die.' His hunch that they were reflective of ancient climate was, however, right and, as his abilities grew, he became a much sought-after and respected figure. He was a character of great colour who for a time was my friend. I admired him much and in return he encouraged me firmly to do many of the things I have done. He was a wildly eccentric and extraordinary lecturer whose party piece on the demise of the Irish elk complete with demonstrations involving helpers, willing or otherwise coerced, to hoist high gigantic antlers for crashing confrontations of their martial prowess was a grand horny triumph.

Russell's knowledge of early zoology was superb. He, for example, knew that early studies involving lynx kittens and scissors conducted at the London Zoo in the late 1800s demonstrated quite graphically that if you cut off their long ear tufts, they became unable to balance and as a result fell out of trees. He knew that if you put several water voles individually in glass aquaria with water and no land, it took them on average ten minutes to drown and that you could not revive a dying porpoise with a bottle of brandy – although it did swim very fast for a terminally short period of time. He had wolf stories galore. Did I know that Scots naturalist Iain Brodie had once offered him wolf cubs, which his wife, Beryl, had refused to entertain on the grounds that:

a) he was the worst trainer of dogs in the world;

b) she loved their children very much and wanted them to live long to have offspring of their own.

Was I aware, he asked once, that centuries ago the Welsh kings, having been obliged after a battlefield defeat to return a tribute of three hundred dead wolves annually to King Edgar, had paid the bill for three years and delivered their carcases to Abbey Dore in Hereford-shire and that, in commemoration of this event, an iron wolf head had been forged on the hinge of a door in the still-surviving church?

Russell died in 2011 and, while his tales could be of the tallest sort, I recalled his mention of the iron wolf and, when my wolf hunt began, resolved to see if this tangible artefact might still be found. Although I knew by then that another memorial of two wooden wolf masks carved into the façade of an ancient house near Glastonbury that were still visible in 1880 had long since vanished, might it be possible that the 'iron wolf' endured?[15]

At the entrance to the village of Abbey Dore is a small, neat dwelling of bold red brick just round the corner from the church. Its name is Hunters Cottage and, although its modern occupants may only deal death to ducks, the focus of those that once lived in an older residence far to its north remain clear to this day. When Walter le Wolfhunte died in Nottinghamshire in 1339, among his possessions was a bovate (15 acres of land) and a house in the village of Mansfield Woodhouse. As Walter's profession was 'chasing wolves outside of the King's forest of Shirewood, if any they found', his home was called Wolf Hunt House and it remains rooted unequivocally to its 700-year-old foundations in the centre of the modern village.[16] If a wolf hunter ever dwelt in Hunters Cottage, his passing is unrecorded and the garden, which embraces it now in a glorious palette of mid-summer colour, no longer affords any air of fatality.

Abbey Dore sleeps in a cultivated land of corn, orchards and neat plantation woodlands. In the Golden Valley, where sheep graze still as they have done for a millennium past, we parked in the yard of a jumbled farm. Bow-backed tin buildings, small cattle sheds, old machinery from the 1960s, nettles and neglect. A burgundy bull with a curly white brow ignored our arrival with a snorting disdain. In the buttercupped orchard, a herd of buxom grey and white heifers of Belgian descent were the sole focus of his attention. They sprang and sniffed around him before dancing beefily away with their tails held high.

My kids ran laughing and screaming up the path through the medieval lychgate and into the graveyard of the church. No embarrassed effort to silence their stridency in hissing tones worked well, but kind passers-by smirked at my failure.

Monoliths of isolated masonry are all that remain now of the great Dore Abbey, which was raised on the Welsh frontier by the Morimond monks who trekked from their homeland of Lorraine. Wrought from local sandstone, their creation stood in splendour for nearly four hundred years before the destruction ordered by Henry Tudor severed the Abbey from its church. Like epidermic burns on the flanks of the Holy Trinity and Mary, its scars though cauterised are stark. The ghost naves and transepts; the lost corner stairs; doors that go nowhere; ribs of arcade arches skeletally exposed to the sky.

Chiffchaffs and dunnocks sang. Collared doves purred from yews that were old but not ancient. At the west entrance in the opposite hedge stood a great oak of over ten feet in diameter with a straight crown of pollard head limbs.

We checked the outside doors, kids calm now. Inside as well where the light through the last tiny panes of Cistercian glass shone in kaleidoscope. There was no iron wolf. We checked again, a grown-up look rather than flippant childish flypast. Still no head.

Had it corroded to mere metal flakes or been thrown aside in the visibly recent replacement of the old church doors? Another archaeologist spoke of how the door itself with its ancient hinged pattern had been copied in the 1950s as a replacement for another in the Tower of London. Perhaps this replicate could be found? But by mid 2020, the world was closing fast for covid. Lockdown left Wales in limbo. Hand-painted boards nailed at angles to roadside fence posts told anyone who cared to read them to go home. Exploring churches or meeting historians was impossible. Although I enquired of the Tower's curatorial staff, no response was ever returned.

By then I had found an image of the head in an old blue pamphlet from the 1920s, grainy and grey but still recognisably complete. A snouty thing with a long muzzle that could have been a dragon or a pig or a wolf. A naturalists' group called the Woolhope Field Club had visited it in 1930 and in a volume of its transactions from 1930–32, one of its members, a Mr A. Watkins FRPS, had taken another photograph that identified the location of the iron wolf amid the double sworls of a cast-iron hinge.

But one wolf can lead to another and, on suggestion, I visited another church nearby.

The village of Kilpeck nestles in the soft hills of Hereford and its church of St Mary and St David is one of the finest surviving examples of the county's medieval school of sculpture. Its honeyed stone stands on a site where worship is believed to have occurred for well over one thousand years. Although a simple structure, it's enchanted by carvings. Fantastical basilisks, angels and green men mingle with whorl-horned rams under its high roof ridge. Startled pigs squeal through wide-opened snouts while fallow bucks with broad antlers lying low along their backs flee approaching huntsmen. A harnessed bear in an accurate halter, done with its dancing days, crunches a teasing urchin pleasingly in its jaws.

The Iron Wolf

It did not matter that the church was closed as the subjects we had travelled to see lined the outsides of its slim corbel pillars. My contact had implied that the 'beaked monsters' I would find there were worthy of examination. Although phallic perhaps when initially viewed in flat form, their eyes with carved central orbits confounded this flippant initial observation. On further examination, they clearly displayed a series of ridges along what might be a muzzle while above, on either side, a similar pattern occurred visibly above their eyes. A few subtle cuts for their short stubby nose and two laid-back ears in outline at the top meant that even without distinct outlines of lines of sharp teeth they looked very wolfish.

For those in side relief, there was no need for subtlety. Standing out visibly in stone with their incisors exposed and lowered fur mantels, a far ear raised and a near ear lowered, they could be nothing else.

Snarling wolves well executed in stone.

CHAPTER TWO

'Here, Here, Here!' Cried the Wolf

The wolf does not concern itself with the opinions of sheep.

Roman proverb

W e live still in a world of wolves and talk about them all the time. Their imagery for good or bad lurks powerfully within the dens of our imagination.

We keep 'wolves at bay', 'wolf down food', are 'wolves in sheep's clothing' or 'keep the wolf from the door'. Men and sometimes women 'wolf whistle'. There are 'big bad wolves'. If you transgress in a high-octane workplace or political circles you might be 'thrown to the wolves'. 'He's got a wolf's coat' means he is imperfect, while to 'walk like a wolf' is to tread stealthily. In France they say, 'I have the hunger of the wolf', or if you are in trouble that 'you have a wolf by the ears'.

'He saw a wolf' is an expression for someone who has a hoarse or rasping voice,[1] and there is an old Welsh saying that 'the wolf's teeth can be removed but his nature can't be changed'.[2] If you complain that an issue that will happen one day has already, and then continue to do so stridently, perhaps your detractors will respond in a manner like the Australian senator Gerard Rennick who, unconvinced of the phenomenon of human-made climate change resulting in global warming, could accuse you of being one of 'The "Climate Scientists" Who Cried Wolf'.[3]

The Roman nickname of 'Lupa' for a prostitute was used for sex sellers because they stole their lovers' wealth; in Elizabethan times, whores were also called she-wolves when they lied or deceived. In Ireland, the term '*sod mac-tire*' meant 'she-wolf-sod' or the 'bitch of a son of the earth'.

Queen Isabella of France, who was the wife of Edward II (reigned 1308–1327), was described by many in her time as a 'she-wolf' as she was believed to have been partly responsible for the death of her husband. Murdered at Berkeley Castle in Gloucestershire on 21 September 1327, the king was held down while a red-hot poker was inserted up his anus.

His screams, it was said, could be heard for miles.

Rapacious tax or money collectors were also considered to be wolfish sorts.

In the eighth century, Archbishop Egbert banned the eating of meat, or its sale for human consumption, that had been cut from a wolf-torn carcase as he believed that their teeth were poisonous. By the 1250s, this ordnance had lapsed and Geoffrey, son of Bernard, quite willingly took two shoulders from a deer killed by wolves in the Forest of Dean. When in 1342, Alan, the Earl of Richmond, offered the monks of Jervaulx in Wensleydale a dispensation to take the meat from any deer they found that had been killed by wolves, his offer was likewise accepted without a qualm.

But extracted wolf teeth were powerful, protective talismans. Smooth, white and sharp, they might be hung on a cord and worn round the neck to ward against evil and ill luck or to protect children from night fears.[4] As late as 1731, in a letter to her son Lord Stafford, Lady Wentworth said, 'I have made your daughter a present of a wolf's tooth. I sent to Ireland for it and set it here in gold. They are very lucky things; for my two first, one did dye, the other bred his very ill, and none of ye rest did, for I had one for all the rest.'[5]

Rubbing wolf teeth against the gums of teething infants was believed to relieve their discomfort and two set in silver in a seventeenth-century possessions list of valuables from a farm at Plas-glas in Wales may perhaps have been used for this purpose.[6]

In Finistèrein, France, newly-weds were encouraged to nail a wolf head to the door of their house to protect them from demons, thieves and dangerous animals.[7] There are stories from Marros in west Wales that wolves' heads were once displayed in the churchyard and Francis Payne reading the Reverend D. Edmondes-Owen's statement about a wolf being killed on the River Edw in Tudor times added that 'his paws were nailed to the door of the church'.[8] As late as 1736, in Widecombe parish in Devon, an instruction was issued that 'whatsoever foxes are hereafter taken or killed within this parish are to be brought to the church and hung up at the parish tree' while in Sedbergh, Yorkshire the 'heads of foxes and other obnoxious creatures used to be nailed to the church door.'[9]

The soft dense winter coats of wolves were valuable trade items.

Though in early Welsh law wolf fur had no legal worth, in later codes it rose in value to the sum of 8 pence, which was on par with the fine soft pelt of an otter.[10] In Ireland at Waterford in 1243, taxes were levied on dried and cured wolf skins, and in the Galway City list of Taxable Goods in 1361 they were also recorded.[11] In 1394–1396, the monks of Whitby were paid 10 shillings and 9 pence for the tanning or 'tawing of 14 wolf skins'[12] and a ship called the *Magdalen* registered 35 skins worth 4 shillings, 4 pence in its cargo list in 1505. Many other imports were recorded through Bristol between 1503 and 1601, with 731 alone being landed in 1558.[13] By the 1570s, they were visibly becoming harder to find. One Alexander Clark, in response to an enquiry from the Countess of Moray, stated that: 'As for the Wolf skins ye wrute for I could get na knowledge of ony at the present.'[14] In 1584, Walter Aikman contracted to buy six hundred wolf skins from Andro Scherar, who was a merchant in

Stirling. When Andro failed to deliver, their legal dispute regarding the transaction remained current when Aikam died in 1589.[15] In 1661, two ounces of silver were paid in Scotland for 'ilk two daker' of wolf skins, a daker being a roll of ten.[16]

At the centre of the Natural Park in the Monts d'Arrée in Brittany lies the sleepy provincial hamlet of Le Cloître-Saint-Thégonnec. The village rests in a landscape where small shrines at the boundaries of obscure settlements are a legacy of a time not so long ago when offerings to the saints might ward against wolves. St Thégonnec, the village's patron, was associated with one. Originally from Wales, Thégonnec was so strong that he tamed and trained a stag to pull a cart to carry stones to the building site of his church. One day a wolf came into the town, attacked the stag and ate it, but Thégonnec, instead of retaliating, befriended the wolf and trained it to pull the cart. When the church was complete, the wolf returned without rancour to the forest while all involved were presumably grateful for its toil.[17]

The village square in Le Cloître-Saint-Thégonnec is occupied today by the Musée du Loup – the museum of the wolf. Outside its entrance, a howling wolf carved from stone is watched over by a wood god carved from dark oak, wearing a cloak made of a wolf skin with its head sewn into a hood. Inside is a riotous ensemble of fascinating artefacts.

Old news sheets and magazines record grand packs of slavering wolves attacking people in all manner of settings.

Dragoons are surrounded; sleighs with hysterical horses overwhelmed in the snow; ordinary folk in fur-lined garb hold them back at arm's length by the throat when they leap through the windows of their houses.

Children's collages of wolves made from bottle tops and coloured paper lead along a corridor into the petite main building where the heads and limbs of Breton wolves killed less than a century ago have been fixed with hand-forged iron nails to old barn doors carefully reshaped to fit into atmospherically ill-lit cabinets. In others, natty dancing costumes with wooded masks, spiky raffia dresses and home-woven tunics pose upright on mannequins. The big wooden clogs that accompany this garb complete an ensemble designed for a 'clickity-clacking' folk dance, which in Breton is called the 'Dañs ar bleiz', that was traditionally practised in town squares to drive wolves away.

Much less elaborately, Irish shepherds once bashed stones together to produce a similar sort of noise, while their Welsh counterparts made rattles from rocks contained in leather pouches, which they slung over their shoulders on a stave.[18] When I cared for Misha and Nadia in the early 2000s, I never thought to test their tolerance of acoustics of this sort. No doubt as a novelty it would have prompted interest followed swiftly thereafter by indifference.

In one of the museum's darker rooms there is a clever graphic thing like a photo booth, which, when you sit inside, transmits your image back in werewolf form. My daughter Maysie looked like one of Paris Hilton's dippier dogs, while my slightly leaner son Kyle looked if not exactly wolfish then more credibly canine. I looked like a zoo wolf. Slouchingly old, overfed and grizzled.

———

There were tribes of people who named themselves after the wolf.

One of the earliest were the East Anglian 'Wuffings' or Wuffingas, which literally translates as the 'kin of the wolf'. Originally from Jutland, they were founded around 575 CE by a leader called 'Wuffa', which is Old English for 'little wolf'. They saw the wolf as their totem animal that functioned as an intermediary or messenger between the world of the living and the underworld of their ancestors. The most illustrious member of the dynasty was King Rædwald, who may have been the warrior interred in the longship at Sutton Hoo surrounded by riches from all over the known world. His hoard contained a gold clasp for a money bag with inset garnets from Sri Lanka depicting the Norse god Woden flanked by the wolves Geri (greedy) and Freki (ravenous) that guarded the borders of his kingdom forever against his enemies, the giants.

The last descendant of the Wuffinga line was King Edmund, who assumed the kingship of East Anglia at the age of around fourteen in 855. He was a devout Christian who was, in his time, considered to be 'the model of good princes ... a declared enemy of flatterers and informers ... The peace and happiness of his people were his whole concern'.[19] In 865, 'a great heathen force' of Danes led by Ivar the Boneless landed in England to form a permanent army of occupation.

When Edmund was captured, he refused to submit unless the Danes adopted Christianity and this Ivar declined to do. On 20 November, the king was seized, tied to a tree, beaten and finally executed with a volley of arrows. His body was then decapitated. When his followers returned to retrieve what remained, they discovered his head being guarded by a large wolf, which cried out in Latin, 'Here, here, here!' Miraculously, when his body and head were reunited, they reconnected leaving only a faint red mark. Whole again, Edmund was interred in the abbey at Bury St Edmunds. In a carved pew end in the church of St Mary and St Lambert at Stonham Aspal in Suffolk, a pug-nosed wolf clasping the king's head firmly between its front paws commemorates this astounding occurrence.

By the thirteenth century, the Wuffingas' view of the wolf had long passed and associations of a very different sort were rising. The physician Rogerius was the first to use the term 'lupus' to describe erosive facial lesions. Although now known to be a chronic autoimmune disease that can damage and disfigure any part of a body's epidermis, joints or internal organs, its past association with the pattern made by a wolf bite led healers to look to wolves for the cure. In 1300, a certain Reverend William from St Margaret's in Ledbury, who was a doctor of some kind, fell foul of customs officers when he tried to import the bodies of 'four putrid wolves'. During the ecclesiastical trial that followed he was denounced as a fraud when he argued that they were required for medical purposes as a cure for 'le loup' (lupus) and could not otherwise be obtained in England.[20]

It was believed that the application of a wolf's severed forefoot would cure pain in a breast or relieve the inflammation in the tusks of domestic boars when they shattered and became difficult to handle. Wearing wolf-skin shoes made children brave, while hanging up their heads kept sorcerers at bay. Wolf hearts or their gall might assuage epilepsy, and their livers, dried and powdered, could cure human complaints of the same. Wolf blood smeared onto a facial scar would remove it completely, while a dried wolf penis made into a ligature and named after a specific man or woman would prevent them from having sex until it was undone.

In the North East of England, 'woof' meant cancer of the stomach, with 'wolf in the wame' being another medieval term for this deadly condition. The lumps caused by breast cancer, open sores or protuberant knobs on the legs were all also called wolves.[21] Among the listed causes of death in London in 1632, ten persons were recorded as having died from 'the wolf'.

Even given this litany of incredible beliefs, it is astounding that sleeping with a severed wolf's head under your pillow was believed to prevent rather than inspire nightmares!

Wolves have long been regarded as greedy and lacking control, attributes illustrated in the ancient fable 'The Priest and the Wolf'. In the tale, a priest attempts to improve a wolf's character by teaching it to read, but when he asks it to name something beginning with 'A', the wolf replies *agnellum*, the Latin for 'lamb', thereby proving its true food-obsessed nature.

It was said that wolves spat on their paws so that their prey would not hear them approaching and that they always stalked their prey from upwind. If they made a mistake by snapping a twig in their stealthy approach, they would punish the paw that had made the error by biting it repeatedly. This allegorical behaviour portrayed in a carved oak pew in Faversham Church is like another at Iffley in Oxfordshire, the only evidence that identifies the grotesquely contorted creature performing the act as a wolf.

The word 'werewolf' or 'lycanthrope' defines a person who genuinely believes that they can shapeshift into the form of a wolf. The prefix 'were' comes from the Old English 'wer', meaning 'man'. In medieval Europe, werewolfery was believed to be real and those who professed to being werewolves were in league with the devil. While it is well attested that some souls deliberately donned wolf skins to help them with their transitions, King James I of England in his 1597 book, *Daemonologie*, was one of the first to advance a belief that victims of this delusion were suffering from melancholia. The English scholar Robert Burton, in 1621, also considered that the men who 'run howling about graves and fields in the night, and will not be persuaded but that they are wolves or some such beast' were suffering from a very definite 'disease'.[22]

Despite this understanding, as late as the seventeenth century those suspected of or confessing under torture to werewolfery might be arrested and executed.

Gerald of Wales related the following tale from Ireland in 1188:

An Irish priest was on his way from Ulster to Meath when he was approached by a wolf who spoke to him. The wolf assured him that there was nothing to fear, it was just that his wife was dying and he wished the last rites for her. They were both victims of a curse pronounced on the people of Ossory by St Natalis, according to which every seven years two people had to don wolf skins and live as wolves. The priest was terrified but followed the wolf into the woods. At some distance he found a female wolf lying ill beneath a tree. The priest stood frozen unable to assist. The wolf reached down and rolled back the female's wolf skin and the priest saw underneath the bony torso of an old woman. No longer afraid, he heard her confession and gave her the last rites. The wolf accompanied the priest back to the high-way, thanked him profusely and told him that if he lived out his seven years, he would search the priest out and thank him properly. The priest went on his way satisfied he had done right.[23]

Scottish folk tales tell of 'Wolf' McAndrews, McKenzies and McGregors who were all removed from their parents and taken away to be raised by wolves. While this practice was once so common for the McDonalds that they were called '*Sliochd a' Mhadaidh Allaidh*' or 'The Race of the Wolf', their penchant for the butchery of more settled clans may also have played a part in gaining this reputation.[24]

It is unclear now if any of the many wolf children encountered worldwide like Dina Sanchar, the 'Wolf Boy of Secundra', who was supposedly Rudyard Kipling's inspiration for Mowgli in *The Jungle Book*, were what they were touted to be. Caught after hunters followed a wolf pack into a cave in the jungles of Uttar Pradesh in 1872, Dina stares back troubled and strange from the images taken of him to thrill tourists. British officials who knew him recalled the easy relationships he formed with street dogs and jackals, but was he really raised by wolves?[25] It's impossible now to be sure but, for the rest of his sad life in the orphanage where he lived, he refused all foodstuffs other than raw meat and only communicated in grunts and barks. Although he could walk upright, Dina preferred to move mainly on his hands and knees. He refused to wear clothes but did learn to drink water from a cup and also to smoke. This unfortunate habit ensured that he came not to the laughing, singing, dancing end devised by Disney but expired instead from tuberculosis in 1895.

In a more modern conundrum Marcos Rodríguez Pantoja, who was born on 7 June 1946, in Añora, Spain, was sold by his parents in 1954 at the age of seven to a local landowner who then passed him over to a goatherd. After the shepherd died he was abandoned and, 11 years later, the Civil Guard found him living in complete isolation from human beings in the company of wolves. When they returned him to civilisation bound and gagged, he howled and bit his captors firmly when he could. Later, in a more nostalgic frame of mind, he recalled this time as being among the happiest years of his life. In 2010, a film titled *Entrelobos* (*Among Wolves*) was made about his experiences.[26]

In 1971 two boys, Colin and Leslie Robson, excavated a pair of small stone human heads in the garden of their house in Hexham, Northumberland. A photograph that was taken of them shortly after the discovery of these orange-sized effigies with stubby noses, an impression of swept-back hair and crude open eyes illustrates that while the mouth of one was shut in a taut, firm line, the other was wide open in what might have been laughter but could well have been a scream. Immediately after they were brought into the Robson home, a series of strange occurrences began. The heads would move when no one was in and bottles would fly across the rooms when they were. Next-door neighbours reported seeing a half-man, half-goat figure in their house and shortly thereafter the heads were given to Anne Ross, who was an expert in Celtic art.

Her own series of bizarre encounters began almost immediately.

On awaking one morning, she saw a hideous apparition:

It was about six feet high, slightly stooping, and it was black, against the white door, and it was half animal and half man. The upper part, I would have said, was a wolf, and the lower part was human and, I would have again said, that it was covered with a kind of black, very dark fur. It went out and I just saw it clearly, and then it disappeared, and something made me run after it, a thing I would not normally have done, but I felt compelled to run after it. I got out of bed and I ran, and I could hear it going down the stairs, then it disappeared towards the back of the house. [27]

Shortly after, her daughter also confronted a large, dark, werewolf-like figure on the stairs, which vaulted the banisters and vanished into a corridor.

Unsurprisingly she was somewhat perturbed.

As soon as the heads were removed from Anne's house the apparition vanished and never returned. In a building blaze of press notoriety, a local man called Desmond Craigie claimed he had made the heads for his daughter in 1956, but the replicas that he constructed to prove his ability were dissimilar in design to the originals.

After passing from expert to expert, none of whom proved to be any the wiser, the heads were eventually sent to Southampton University for date authentication.

They passed them to a local psychic in 1978.

Who disappeared quite utterly shortly thereafter without trace.

On 30 May 2021, the *Daily Star* ran with the illuminating headline, 'Werewolf Hunter Reveals UK's Paranormal Triangle Where Terrifying Beasts Gather'. The story was based around the sinister location of the 'Wold Newton Triangle' in Yorkshire, which zombies, fairies and dragons have known about for some time. A phantom river called the 'Waters of Woe' flows through its centre and a rough, hairy beast called Old Stinker with a terrible halitosis acquired from the consumption of decaying human corpses roams its banks. Some accounts say that this creature has red eyes and Charles Christian, the werewolf hunter who first raised concerns, recalled that, 'When I was a child, I remember someone saying they would not drive along the road from Flixton to Bridlington after dark because … when people would glimpse what they thought was the rear lights of a car in front, it would instead reveal itself to be the red eyes of a wolf.'[28]

Sir David Attenborough has just booked a film crew to produce his next full series at this convincing location.

CHAPTER THREE

The Howling Gods

I see the ridge of hinds, the steep of the sloping glen,
The wood of cuckoos at its foot,
The blue height of a thousand pines,
Of wolves, and roes, and elks.

Domhnull Mac Fhionnlaidh, 'The Aged Bard's Wish'

The four north-facing Bone Caves of Inchnadamph in Sutherland look out from the base of the limestone cliff of Creag nan Uamh over the glen of Allt nan Uamh. At its head, the ancient calving grounds of the reindeer lie blessed by the spring sun on the rising slopes of Breabag. On long warm days, after a good kill the wolves must have lain lazy and fat. While some slept soundly, their dark-coated cubs above ground would have clumsily tumbled and bounced on pestered parents at rest. Perhaps they stalked petite pikas singing on the rocks nearby, creeping towards them ineptly on paws that were too big for their task? Did the ravens patrolling the sky above ever glance down at their game?

In the midst of the grunting grazing herds, the life pattern of the wolves that lived there would have flowed in an annual rhythm of opportunities, matings, bondings, births and deaths for millennia.

Until one day it stopped.

After dwindling for centuries as the earth warmed, the snow melted in season and the trees grew faster and taller with every passing year, the ice beasts that once called Britain home faded perhaps eleven thousand years ago. The wolves that carried the antlers and bones of the reindeer down to gnaw in the cave depths would have been among the last to leave and the remains of one of their last litters of cubs have recently been found. Although no carbon dating of their remains has so far defined their time of life and death, they were too small to leave of their own accord and, when their mother failed one day to return, they would have perished, whimpering, cold and alone. Maybe by the time of their birth she was solitary too. One of the last, with no pack mates to help or to care for her offspring as wolves often do.[1]

The commonest species of wolf worldwide that still inhabits a huge swathe of the northern hemisphere and was once found in Britain is the grey wolf. There are wolves in the eastern forests of Kamchatka, Russia, that lap the Pacific, and wolves have returned to inhabit the cathedrals of gigantic, moss-laden conifers that mantel the peaks of Oregon's Cascade mountains on the ocean's west flank. Their forms vary greatly and the large white wolves in Canadian Labrador and petite fawn ones that abide in the sandscapes of Arabia are all in reality of the grey sort. While a good-sized male wolf in the north can be over six feet in length and weigh around sixty-five kilogrammes, in the south a substantial desert wolf might only be twenty kilogrammes. Females are around 20 per cent lighter than their mates.

Some creatures called wolf are not in fact wolves at all. There are long-legged, maned wolves in South America whose nearest relatives are bush dogs or forest foxes; an aardwolf in southern Africa that is an insectivorous hyena and, in the alpine regions of the Bale Mountains in Ethiopia, a brick-red wolf (*Canis simensis*) that, while related to its cousins in Europe, is in form and function closest in type to a jackal.

There are wolf fish and wolf spiders and, while these creatures are named for their hunting abilities, the large russet field hamster where it occurs in the Netherlands is called the *korenwolf* or 'grain wolf' as a result of its penchant for seed gathering. In Britain, the common otter was formerly called the river wolf as it stole 'our' fish, pike were 'fresh-water wolves' and the orca was known as the 'wolf of the sea'.[2]

In short, everything that took from us in large or small consequence could be readily described as a wolf and as a result be hated.

While wolves cohabit with jackals throughout much of their Eurasian range, hybrids between the two are relatively uncommon. Coyotes (*Canis latrans*) and wolves however share a common ancestor and, although their division occurred nearly 1.5 million years ago, the two species have never entirely forsaken a lingering love. Modern genetic studies identify that the eastern wolf (*Canis lupus lycaon*), which is very similar in physical appearance to the grey wolf and was only recognised as a distinct species in the early twenty-first century, has, like its smaller cousin the red wolf (*Canis rufus*), hybridised extensively with coyotes.[3]

So thoroughly has it enjoyed the experience that the eastern wolf is now believed to be a mixture of coyote and wolf in roughly equal proportions. The dainty red wolf, which was once widespread throughout the Southern United States, has been so additionally diminished by habitat destruction, pitiless predator-control programmes and the loss of its natural prey that hybridising with coyotes was its only option when pure mates became impossible to find. By the late 1960s, a mere seventeen remained wild in Louisiana and Texas and, when these were captured to begin a conservation breeding programme, only fourteen proved pure enough to use. While a reintroduction project using captive-bred individuals, which began in 1987, saw the wild-living red wolf population rise to a peak of well over one hundred free-living individuals, many of the old problems remained. In 2015, the United States Fish and Wildlife Service declared that it would no longer sanction the release of any more captive-bred wolves into the wild nor continue a sterilisation for wild coyotes to protect their genome. Although this decision was subsequently overturned by court action enabling the release pro-gramme to be recommenced, any vision of the red wolf surviving independently of active intervention remains elusive.[4]

In the Southwestern United States, a similar project to secure the future of another small wolf may have a happier ending. While the Mexican wolf (*Canis lupus baileyi*), which was also once known as the 'lobo', is a subspecies of the grey wolf rather than a distinct species, it is the smallest form of wolf in North America. Its coat is generally dark along its back and tail, while its flanks are greyish yellow. Once worshipped by the Aztecs, by the mid-1970s overhunting, trapping and poisoning ensured that only a small number in captivity and a few in the wild remained. In 1976, the Mexican and United States governments collaborated to capture the remain-ing wild wolves. Four males and one pregnant female were caught between 1977 and 1980 and used to start a captive-breeding programme. The progeny from this were released into Arizona and New Mexico in 1998, and by 2022 surveys indicated that there were over 241 wild Mexican wolves, while more than 350 remained in the captive-breeding programme.[5] Approximately 60 per cent of the population currently occurs in New Mexico and 40 per cent in Arizona.

Though recent genetic studies have identified some coyote genes in the free-living lobos, it seems that this form, for whatever reason, has not integrated with them thoroughly and as a result they remain largely wolf.

Captive-bred pups are still being integrated into the dens of wild-born litters to broaden their gene base and the future of the Mexican wolf looks promising.

We do not really know what British wolves looked like. Some may have been dark. In Northamptonshire the village of Blakesley is derived from '*Blaecwulfs lēah*' ('black wolf's meadow' in Old English) and, although the stories about black wolves being killed in Britain have been dismissed as fanciful, they exist widely in North America where recent studies demonstrate that this colour form is directly derived from an ancient hybridisation with dogs.[6]

MacQueen's adversary on the Findhorn could easily have been of this sort.

While a few of the skulls found in the depths of the limestone caverns in the north of England suggest that they were large, perhaps only a single description of wolves on these islands that is not based on wild exaggeration exists. It comes from Ireland and recalls that 'the talk ran of wolves and of those that showed the best sport as the large grey wolf, and also the smaller dun one, which two it would seem are the sorts which chiefly prevail in this country.'[7] While wolf cubs can be much lighter in colour than their parents and it may be that this reference relates to nothing more than this natural difference, the Irish wolf expert Dr Kieran Hickey has long argued that the Irish wolves may have been a subspecific form or even a species unique in their own right.

It is unlikely now that we will ever know, as techniques for effective taxidermy did not develop until a time when wolves in Britain were scarce. Perhaps only a single specimen killed by Sir Ewen Cameron of Lochiel in 1680 may have survived into modern times, while the set of dried wolf paws from southern Wales that are lodged now in Cardiff Museum are bare and hairless. Cameron's wolf was sold as 'a noble animal in a large glass case' at auction in London in 1818. Purchased by Sir Ashton Lever, who commissioned its lithographic illustration, the image that remains implies that its stuffer was of questionable sanity.[8] Residing in a glass case in front of a mountain montage, the paws of its protruding forelimbs turn inwards awkwardly on their sides. Its face, with ears bolt upright and large glass goggle eyes, has a startled expression while its mouth, opened in a dislocated jaw sort of sense, mimics fairly the sort of death inspired by anal electrocution.

It was a truly memorable exhibit. Old, worn and lost.

I enquired after others.

One day I found myself standing in the grounds of Wolvesacre Hall next to a boulder the size of a Land Rover chatting to James Hall, the dapper land agent for the Cholmondeley Estate, and two friends, Michelle and John Clarke. It was a sunny afternoon and yet another herd of perplexed cows of the black-and-white sort were viewing us closely with interest. John had just explained that the boulder was supposed to be a marker of the place where the last wolf in Shropshire was killed when

Michelle asked if he 'had shown him the photograph'. After confessing he had forgotten, John pulled out his mobile phone and produced the oddest of images. At one side a Victorian gamekeeper, grey-bearded in tweed, stands behind a woman sitting in a long black dress, while on the other a younger man with a black moustache and flat cap accompanies a younger woman in a similar position. As was the fashion of the time all look solemn in sepia while, on the table between them mounted on a shield-shaped plaque, was the stuffed head of an animal.

Sad and distorted with bulging glass eyes, it appeared too big for a fox and too small for anything much else. On its back in pencil a statement said: 'The last wolf to be shot in the UK at Wolvesacre, the Wych, Malpas. Date Unknown. The Crew Family Gamekeeper to the Godsall Family.'

I sent the image to Dr Andrew Kitchener from the National Museum of Scotland, who has stuffed more living creatures than any other individual on earth – some in his own time for fun – who opined that, 'Given the long ears I would have said this is more likely to be a coyote than a golden jackal, but it's hard to be sure given the poor taxidermy and the poor photograph. I guess a coyote … it's a bit more like a wolf as well.'

In Belleek Castle in County Mayo, a last claimed wolf killed in Ireland at a time unknown still stares back at visitors from within its glass case. No DNA has ever been taken from this whole mounted specimen and no radiocarbon date proves its age, but at least it has a good tale attached. Found by an eclectic collector of antiquities called Marshall Doran, an ex-merchant navy officer who was on a driving holiday to a place called Maam Cross in 1970, it was purchased after a long day in the pub from a local man who assured him blearily that it was 'an object of great age'. When Marshall drove his cabriolet unsteadily home, his new acquisition perched on an open back seat caused great consternation when it was perceived as being alive in the sheep country of Connemara.

Undaunted, the wolf survived its journey back to the castle, where it remains to this day.

———

Grey wolves were domesticated in northern Eurasia between fourteen thousand and twenty-nine thousand years ago and are still considered to be the principal ancestor of the domestic dog. They are great travellers that are well capable of covering vast distances with ease, with some individuals having been recorded as moving over five hundred miles. Where they live in well-organised packs, they can kill large herbivores such as bison or moose, and specialise in ability. In Voyageurs National Park in northern Minnesota, beavers constitute 42 per cent of wolf diets between April and October. Researchers there have identified that on average a single wolf in Voyageurs can consume between six and eight beavers a year.[9] Elsewhere, in Alaska other wolves swim short distances out to sea to access offshore rock formations to predate bewhiskered sea otters that once believed they could lazily bask in peace.[10]

Despite this prowess, wolves will readily consume small mammals, rabbits, amphibians, fish, birds and even insects. When the brambles were ripe in Kent, my cubs would take a lip-smacking delight in the consumption of the dark purple, ripe berries when they were sugar rich. Consumed in bulk, they turn poo bright purple (if you don't believe me, you can prove it by using yourself, your kids or your parents as guinea pigs) and my wolves were no exception. Although the ripe soft ones were a treat, the bitter red ones pruned by accident by their small front teeth would be met with a grimace before being swiftly spat out or wiped from their tongues with a forepaw.

Given their broad dietary spread, it is entirely pre-dictable that British wolves would have consumed other predators that we now consider to be sizeable. A few years ago at dusk in Poland, I was out in a forest in the Masurian Lake region with a warden trying to observe elk. He informed me that, a few weeks before in the very same spot, he had been surprised to see a pair of unusual mammals with black-and-white striped faces. Thinking that these might be raccoon dogs, which I had never seen before, I extracted a field guide from my backpack to agree their English name. When I showed him their picture, he said no, before eventually identifying the creatures he had observed as badgers. When I told him how common these were in Britain, in vast setts of great age dug well back into dry woodland banks, he said that they could never live like that in Poland.

When I asked why, he replied 'wolves' and, thinking back, this makes perfect sense. If wolves lived in Britain, they would ambush badgers on their well-worn territorial paths that they use for evening ambles and, if they could, crunch them up completely while they slept in their snuggly beds. While the badgers' relaxed, sedentary lifestyle would revert to a rollercoaster ride from hell, wolves are also partial to foxes, and any otters planning the long-distance treks over mountains to view new river territories might also be in for a nasty surprise. These adjustments in dominion for minor predators are what happens when a keystone king returns to claim its throne.

While wolves can live well into their teens in captivity, any wolf over ten years of age in the wild is old. Large packs of up to two dozen individuals have been recorded although those numbering between six and ten are more common. Where severe persecution has occurred in the past, or if they start to colonise new ranges, single individuals or pairs can be the norm for a time. Packs that comprise an adult breeding pair together with their attendant offspring of various ages have members with distinct characters and strong social bonds can form between different personalities. The alpha pair assert continuous dominance over their subordinates and, in doing so, dictate the activities of the pack. Although the alpha female principally takes care of her cubs, many of the other pack members will bring food that they regurgitate for her at the den. As the cubs grow stronger, adult wolves who are not their parents will bring food to them directly or play with them when their mother is absent. A wolf pack's territory varies in extent depending on both habitat quality and prey abundance and, once formed, a territorial pack will defend whatever landscape it occupies vigorously against neighbouring packs.

Wolves communicate with one another by visual signal, scent marking and howls that enable individuals to remain in contact and strengthen social bonds. They breed between February and April with around five to six dark chocolate, blind cubs being born in an underground den after a gestation period of around two months. After a few weeks, the pups are usually moved from their place of birth to a location where they can play above ground while the adults hunt. They grow rapidly and are moved further and more often as the summer progresses until, with the onset of autumn, they begin to follow their pack through its wider territory. After two or more years, some move away to find their own mates or establish new territories.

Large ungulates such as elk and wild boar will attack wolf cubs when they encounter them above ground. Given the deadly determination my Heck cattle expressed in a tails up, horns down, bellowing rampage whenever my sheepdogs entered their paddocks, I cannot imagine that aurochs and wisent would not in the past have done the same. One study in the Naliboki Forest of Belarus identified that large male lynx will readily enter wolf dens and force the females out in order to kill their entire litters of cubs. Between 2012 and 2016, when the lynx population began to rise, 63 per cent of the wolf pups produced were killed by them, and between 2016 and 2017, as lynx became common, this figure rose to 96 per cent. In the end, their predation by lynx became so thorough that wolves only survived in the study site as a result of other subadults' immigration from elsewhere.[11]

Nature is a great leveller of balance.

———

In 1897, a grainy black-and-white photograph was taken of an unassuming grey beast standing quite calmly on a chain in the back yard of a property in Sicily. It must have been well tamed as other old images taken of chained wolves in people's back yards demonstrate either snarling or abjectly cowering individuals. It had moved slightly when the photographic plate was exposed and so the right-hand side of its face, turned quizzically towards the camera, is blurred as a result. Though this distortion invests the only photo that exists of a live Sicilian wolf with the kind of air perpetrated by the Victorian pedlars of faked after-life images, the subject it depicts was once real enough. Although Sicilian wolves clung to existence on their island of tightly packed small farms in a landscape largely flayed bare by the Romans for two thousand years after their departure, the last confirmed sighting of one occurred near Palermo in 1924 and while they may have lingered into the 1970s it is now certain that this 'lion tawny' wolf has gone.[12]

There are no photographs of the Icelandic wolves, which were expunged ten centuries ago, nor the cinnamon wolves of the tall conifer forests of the Cascade Mountains on the North American west coast. The black wolves of the Floridian swamps, the white wolves with grey crests down their spine from Newfoundland, the wolves from the Mogollon Mountains in New Mexico, from the Great Plains of Texas, and the Kenai Peninsula in Alaska have all departed leaving barely a trace of their being.

While most of the foregoing were again simply distinctive races of the grey wolf, this was not the case elsewhere. In early 1905, a 25-year-old American traveller named Malcolm Anderson was in Japan collecting animal specimens for the London Zoo and the Museum of Natural History under the patronage of the Duke of Bedford. On the morning of 23 January, three hunters who had heard of his mission brought him the body of a wolf. When he casually purchased its corpse and sent it home together with a vast collection of other specimens, Anderson had no idea that it would be the last relic of the Honshu wolf the world would ever see.[13]

Preserved in a drawer of the Natural History Museum, its dried fur with creamy-tan roots is dark-ash at its tips. When it lived and breathed, its tawny pelage earned it the title of the 'golden wolf'. One of its rich amber-tangerine eyes survives desiccated and dry. Where the other eye once was, a sisal cord laced through the empty socket secures a worn cardboard label. Measuring only thirty centimetres at the shoulder with a total body length of about eighty-eight centimetres from its snout to tail tip, this wolf was the smallest ever known. Modern studies suggest that it may also have been a significant contributor to the development of the domestic dog. If its unique characteristics afford the Honshu wolf in time the status of a distinct species, only five taxidermy specimens survive in the world's museums to accept this accolade.

In Japanese folklore, the Honshu wolf was considered a forest spirit and its presence was honoured with shrines. Many of these still exist and place names such as Okamiiwa, 'Wolf Rock', and Okamitaira, 'Wolf Plateau', recall their presence. Farmers viewed them with reverence as the protectors of their crops against the ravages of deer and wild boar.

Though archaeological evidence indicates that the Honshu wolf once roamed three of Japan's main islands (Kyushu, Shikoku and Honshu), it was not found on Hokkaido.

That was the realm of the Ezō wolf.

This larger wolf, which weighed in at around eighty pounds and stood around eighty centimetres at the withers, was light grey in colour. American agricultural advisor Edwin Dun described it as 'a formidable beast but not dangerous to man as long as other prey is to be had for the killing'. He observed that its 'feet are remarkable for their size, three or four times larger than ... the largest dog which they resemble in shape'. This characteristic, he noted, enabled 'them to travel rapidly over deep snow that ... tires a fleeing deer'.[14]

The Ainu people revered this wolf as their 'howling god'. Ainu hunters would leave portions of their kills for wolves and believed that they could share a wolf's kill if they politely cleared their throats in its presence. In 1701, the first bounty for their killing was introduced on Hokkaido and by 1742 professional wolf hunters were using firearms and poison. In 1868, the Emperor ordered the modernisation of agriculture and Edwin Dun was employed to direct a change from rice cultivation towards ranching. He believed that as wolf predation was inhibiting the propagation of horses in the south-east of the island they must be eliminated and the government dutifully declared them to be 'noxious animals'. Entrusted with their extermination, Dun promptly instigated a regime of mass poisoning involving the use of strychnine baits. When they dwindled in rapid response, bounties secured the last.[15]

In less than a generation the 'howling gods' were gone.

— CHAPTER FOUR —

There I Saw the Grey Wolf Gaping

Isabelle Le Deuff from Coataven Mill, aged 9,
was devoured by an animal said to be a privately owned wolf,
at around six in the evening on 13 August 1773.

R.J. Le Siner, priest of Cadol

At Palacerigg Country Park in Cumbernauld, Scotland, we kept a herd of white park cattle. These striking creatures with their black noses, ears, hooves and long horns had been a feature of the ancient hunting forest that once surrounded the now grotty town of Cumbernauld. They were quite docile creatures, well used to slowly following a bag of cattle cake from one paddock to another.

One night long after dark a knock on the door of my cabin informed me that they had somehow wandered out of their field. As the central part of the country park contained a golf course with neatly trimmed fairways, bunkers and pools, of which the local authority was proud, I thought it prudent to secure them swiftly so, after collecting feed to entice them to follow, I went out to see where they were.

I found them grazing at the edge of a track in full, blue moonlight under a sky sprinkled with the brightest of stars. They came to me when I called and we began to move without issue towards a nearby field. It was late. I was tired. The gambolling calves raced out in front of their mums and the rhythmic plodding of the big bulls' hooves was accompanied by a soft snorting symphony from the rest of the herd.

Lulled into a rhythm of content, I forgot that our route would bypass the large pinewood where our pack of North American timber wolves lurked. While I may have forgotten them, they sure as hell knew we were coming. As we passed at a distance of around fifty metres, they let rip. No howling but a snarling, snapping crescendo. Instant and sudden in intensity. The cattle stampeded, tails and heads high, their hooves gouging deep into the soil. The calves ran flat out while their mothers followed, udders flailing in a bellowing array. While the wolves screamed and jumped against the mesh of the fence, I could hear but see nothing at all. Among the wooded understory they were invisible, hidden by the night. Had they been unrestrained and able to sally forth into pursuit their timing would have been impeccable.

At the bottom of a slope the cattle, blowing gouts of steaming breath from their broad flared nostrils, slid, circled and gathered into a tight formation. Wide-eyed, they waited until I pantingly caught up. Amazingly, although startled and nervous, they then followed me the short remaining distance into a field, where I secured them for the night. The damage they had done to what turned out to be, in the light of day, a once-manicured putting green was impressive. With its limp broken flag and shit-splattered surface it looked like some sort of minor battle had been fought and lost on an obscure frontier.

Though the small beardy greenkeeper was lost for words in the morning, his patrons, with their diamond jumpers, ridiculous trousers and pantomime shoes, swore a lot at my boss.

———

Tame wolves can be dangerous and, as a species that is behaviourally programmed to rise up through a pack structure to a position of dominance if, in their mind, you are part of that pack, you can become the only obstacle to their desire.

Edward of York was an English noble and fanatical huntsman who published his book *The Master of Game* in 1413. He cautioned that:

Men cannot nurture a wolf, though he be taken ever so young and chastised and beaten and held under discipline, for he will always do harm, if he hath time and place for to do it.... For he knoweth well that he doth evil, and therefore men ascrieth and hunteth and slayeth him. And yet for all that he may not leave his evil nature.[1]

But the Irish kings had long kept tamed wolves. As a totem of power or savagery subdued, there would have been no better emblem and the early Irish 'Brehon' laws dating back to the seventh century make it clear that any offences committed by a tame wolf were to be charged at the same rate as those for a domestic dog. They understood rationally, as we have forgotten or chosen to ignore, that the damage that both can inflict is the same. Transgressions included attacks on humans and domestic animals, though, when a domestic dog and a wolf teamed up to kill sheep together, deciding an appropriate penalty was more complicated.[2]

There is evidence that wolves were on occasion maintained by the Welsh. During the second half of the fifteenth century, a drover, woollen merchant and poet called Tudur Penllyn lived in a farm near Bala. His son, Ieuan, who was also a poet, clarified in prose that he had no brothers or sisters before going on to imply that a wolf had castrated his father and that it was an '*Iledfegin*' or domesticated wild animal.

No sisters to play
Nor brothers for me
No seeds of possession be
It can't be for there are no testes.[3]

There I Saw the Grey Wolf Gaping

In my own time, I have been attacked twice by wolves. On both occasions, hand-reared, captive individuals were involved. The first was a trivial and recurring event, much enjoyed by the wolf, while the second was a stupid mistake caused entirely by my own foolish actions.

Between 1989 and 1994, I managed a collection of European wildlife species at Palacerigg Country Park. Its original director, David Stephen, was an unusual character. Born in Airdrie into a poor working-class environment, he knew very well what poverty looked like and as a boy escaped whenever he could to the countryside. The first part of his working life was spent as a Poor Law officer in Glasgow, but he turned to full-time journalism after the Second World War, firstly with the *Daily Record*, where he popularised nature to an extent never known, and then for the *Scotsman*. He was a self-made man. Big and brash but very knowledgeable with a toughness gained from his early days that he never lost. David had retired before I was employed and on occasion, as he still lived in the manager's house, I would go round in the evenings for a chat.

David had a tame male wolf obtained from a seedy zoo in Aberdeen, which he called Marquis. Though it adored him, it hated almost everyone else. It was a bastard.

The central part of the country park contained a series of large carnivore cages and a children's farm. As a requirement of zoo legislation, the wolf pen had a tall stand-off barrier of rigid post and rail fencing covered with one-inch-squared sheets of welded mesh between the footpath, where people viewed Marquis and his Russian wolf-wife, Magda (who spent most of her time sleeping), and the perimeter chain-linked mesh of their enclosure. While the distance between the barrier and the enclosure mesh was well over arm's length – to prevent very stupid people, of whom there were many, sticking their hands through – it was nevertheless sufficiently wide to function as a perfect trap for windblown litter. As a result, on a weekly basis this void had to be cleared by working in teams of two with one jumping the fence to litter pick and the other keeping watch.

Now, chain-link wire mesh for those of you who know little about fencing is hexagonal in shape. It is knitted into strands by giant machines and then sent forth for erection in large round heavy rolls. It can last if galvanised for very many decades and it is commonly used to enclose military bases. Its only slight disadvantage is that if something hits it hard enough it responds in the same manner as the base of a trampoline when bounced on. This quality, when bending down to pick up litter, between a position of great give and one of none at all, ensures complete vulnerability.

Marquis knew and relished that knowledge.

And so, I would begin with the chewing gum, the tiny see-through cellophane wrappers from toffees or fudge, the cigarettes, the Styrofoam coffee cups, the used nappies, on occasion an odd syringe, maybe a condom or two and, as it was central Scotland, discarded bottles of the national drink, Buckfast Tonic Wine. If I became distracted or immersed in this task, if my watcher got bored or thought it was funny, the waiting wolf would hit the fence like a juggernaut and the fence would give as if constructed from rubber. His tiny sharp front teeth set between gigantic incisors, which, thank all that was holy, could just not quite fit through, would snap shut like steel castanets tight on my butt, leaving a neat hole in my trousers and causing a bruise the size of a baseball to form swiftly thereafter.

It was impossible to sit down easily for days thereafter, while regaining full control of my colon function took much, much longer.

It was a distinctly unpleasant task.

The wolf, fun over, would yawningly return to snooze.

I am no wolf biologist and will not pretend to understand the species well. In a lifetime working with many creatures, you get to know some better than others. I have always preferred the cuddly beaverish ones that gnaw.

In Kent I managed a much larger zoo, which had a group of European wolves that had been rescued from a Romanian fur farm. Kept in a large woodland enclosure, the initial group consisted of a rather shy male and two females, which were neither overly tame nor timid. At that time, there was a something of a vogue for hand-rearing tame 'wild' creatures for educational talks and presentations.

Looking back at this trend, I am no longer convinced that it was anything more than an extension of the tawdry novelty identified by John Caius, who remarked in 1576 that the only wolves left in England were those 'brought over from beyond the seas for greedinesse of gaine, and to make monie onlie by the gasing and gaping of our people upon them, who covet oft to see them, being strange beasts in their eies', but at the time it seemed valid.[4] To tame the wolves, you had to obtain wolf cubs before their eyes opened. To secure the cubs, you had to go into the enclosure armed with brooms and investigate its earthen banks for deep cubbing dens, at the bottom of which you would find a dark squirming litter.

The first year we did this passed without significant issue. The following year the mother wolf knew what was up. Her keepers had reported a perceptible change in her behaviour after birthing – stalking towards them hackles up and growling when they fed her from the gate – so we considered it prudent to organise. Based on the finest dictums of the Roman army, we decided to form a kind of Roman testudo or tortoise formation. We made shields of our own height from rigid hexagonal wire mesh and fitted wooden handles to their insides. After deciding that the creation of crested helmets, while a tempting final flourish, was unnecessary, our plan was that four of us, large and male, would enter the enclosure in formation with a sturdy yard broom apiece in a tight square and work our way towards the den. Three would stand over its entrance while the other extricated the cubs.

There I Saw the Grey Wolf Gaping

What could possibly go wrong?

The formation worked well and we got in OK, but the mother wolf was by now thoroughly disgruntled and began rushing straight for us, trying hard to bite our ankles under the shields. While we held firm formation with our shields on the ground, she had no opportunity to do any damage and, although obviously displeased, she drew back for a time.

We checked the first bank with both female wolves now circling. No cubs in the dens. The second bank produced the same result. In the third, there was a tiny hole and, when I crouched down to shine a torch inside, I could see a writhing litter of small, blind dark brown cubs in its base. The problem was that, however hard I tried, I couldn't fit in. Poised over the entrance in a shield huddle, we conferred. If we moved and their mother went down into the den it was game over without any realistic possibility of a repeat performance.

I should have stopped there.

Instead, I did not and, after checking just how far we were into the wolf pen from the entrance gate, which we had latched shut on the inside, I called Daisy, one of our small, slim keepers, on the radio for help. I turned to my colleagues and said that I would move with my brush and shield to let her in when she came while they waited where they were at the den.

Daisy duly arrived and I moved out of formation to get her. The distance from the gate to the den was no more than the length of a medium kitchen. I could see into the cleared ground in front of the gate. No wolves there. I lifted my brush free, grabbed my shield firmly and turned towards the gate. I took I think perhaps three or maybe four steps. I am not sure. In life when bad things occur, they generally do so in slow motion. I was no distance from Daisy when I saw her eyes widen and heard somebody behind me scream.

As I turned back towards my colleagues on the mound the mother wolf came towards me at speed. She leapt for my shield and, as I swung my brush round to ward her away, she turned in an instant and snapped its head clean off. I did not immediately see what happened next as Daisy shouted that another wolf was coming. I turned holding what was by then a very short pole to see the second female racing in from the side and, as I did so, tripped backwards over a tree root behind me. I remembered thinking as I fell that it was all going to hurt very much before I hit the ground.

But no wolves. No tearing horror. I sat up. The chaps in their triangle were still breathlessly there looking shaken. Daisy, white faced, was clinging on to the gate. The wolves were nowhere in sight but the brush pole I held was splintered and short. I stood up and got Daisy. She slithered swiftly down into the dark, got the two cubs we required and we all got out.

I felt shaken and sick.

It transpired that the first female had, having severed the brush head from the shaft, turned neatly and run with it back into the woodland. The second female, as

I fell, swerved to follow her and, within sight of those watching, the two wolves had pinned the broom head firm with their paws and stripped it ragingly bare of bristles with great bites.

It was a bloody lucky escape.

But a pointlessly stupid thing to have done. I have never again attempted anything of the like and even though the cubs we took, Nadia and Mishka, grew in the end into great characters, it was a witless undertaking. It did however demonstrate the speed, power, determination and coordinated ability of wolves to attack if they choose to do so. While my badly armed, wooden-broomed vulnerability was near comedic, if you had been a lone traveller in a dark medieval past with a sword, staff or flintlock you would have stood little in the way of a chance.

Although the Saxon description of January as '*Wulf Monat*' was considered to refer to a time when the starving packs were at their most dangerous, it may have been the fear of diminishing winter provisions that really bit deep. But the Gaels called the last fortnight of winter and the first of spring the '*Faoilteach*' or '*Am Faoilleach*', meaning 'the time of wolf ravaging', and other folk tales also speak of their threat.

The International Wolf Center in Minnesota identifies that wolf attacks almost always stem from familiarity with humans, provocation when they are trapped or as a result of great human alterations in natural environments, or rabies. The IWC goes on to state that, 'A person in wolf country has a greater chance of being killed by a dog, lightning, bee sting or a car collision with a deer than being injured by a wolf.'[5] All the criteria identified by the IWC that could trigger attacks would have historically been drearily prevalent in Britain. Given that the often-indivisible propaganda of church and state was in full swing against them, any battlefields scavenged by wolves, corpses disinterred or chorused howls in the dark would simply have fuelled the flames of dread.

Red-Riding-Hood-type tales told by the Brothers Grimm and others preyed on this by merging an ensemble of dangerous strangers with dark forests, villains and saviours. When wolves became rare in settled landscapes, their continued existence in remoter parts filled travellers unfamiliar with their presence with apprehension. In 1618, John Taylor the Water Poet travelled on foot to London. When near Braemar, he stated, 'I was in the space of twelve days before I saw either house, cornfield, or habitation of any creature, but deer, wild horses, wolves and such like creatures, which made me doubt that I should never have seen a house again.'[6] If Teddy Roosevelt only observed in the late nineteenth century that the wolf was 'the archetype of ravin, the beast of waste and desolation', his view would have been shared by most Europeans for a millennium.[7]

But this dread based on fact or fiction?

Once upon a time near the farm of Malling, to the west of the Lake of Menteith in Perthshire, a girl was 'attacked' by a wolf when she was carrying food to the harvesters. The details are sketchy but basically, as the beast emerged from the forest of Menteith, the girl dropped her basket and ran away, leaving the wolf to consume its contents rather than her. Another story has been spun around a location called the 'wolf's bush' near Ardross Castle. An old woman going to get a griddle from a neighbour stopped to relieve herself in the thicket and on her return saw a wolf scraping the ground where she had been. She hit it with the griddle and ran back to her neighbour's house. They responded promptly with every destructive weapon they could find and beat the injured beast to death.

While this affords another griddle-girdling account, neither speak much of attack.

In AD 937, King Athelstan granted the building of a hostel at a place called Spital Ho near Saxton Hill to afford travellers shelter from wolf attacks. But the erection of facilities of this sort, which was undertaken by Achorne, Lord of Flixton, to retain 'one alderman and fourteen brothers and sisters of that place for the preservation of people travelling that they might not be devoured by wolves and other wild beasts', also dented one's purse.[8] As late as 1577, it was noted in *Holinshed's Chronicles* that wolves remained so dangerous in the time of the doomed Mary Queen of Scots that it became necessary to erect more overnight refuges or spittals to ensure the safety of Highland travellers. The Spittal of Glenshee, on the Devil's Elbow Road from Blairgowrie to Braemar, is but one remaining example.

———

We took the cubs for walks in the forest that surrounded the wildlife park when we could. As long as you had one robust hand-held lead and a stout chain clipped to your belt, which, at worst, would ensure that if they ever pulled you over by accident your body would painfully act as an anchor, then all was just fine. They enjoyed their outings very much.

Unfazed at all by people who moved towards them in an open manner, they would greet those who did with ease. Complications came not with the occasional deer passing across the tracks, which while interesting was worthy of no rapid response, but from an altogether swifter prey that haunted the forest paths. Commonly these creatures wore spandex in a variety of lurid shades and odd shiny helmets turned backwards. When they pedalled towards us sweating down the long tracks it gave the wolves full chance to prepare. Although they might strain mightily forward, we knew what was coming and, by gripping a tree with one arm, we could hold firm. When the same prey however, on its strange, squeaking wheels, more slowly traversed the narrow muddy paths that led through the dense sweet chestnut coppice that lined the main track, their sudden appearance made a clean ambush by the cubs more likely.

The American colonist Thomas Morton observed in 1637 that his actual experience of wolves was that they were 'fearefull Curres, and will runne away from a man ... as fast as any fearefull dogge.'[9] In 1853 Jean Guillot, a wolf hunter in Brittany who sought out their dens in order to capture wolf cubs for multifarious purposes, would simply take a stick and a basket for this task. Although their mother would commonly follow him for a time through the forest calling to her whimpering pups, he was quite clear that never once had he been attacked or ever even felt threatened.[10]

In modern times Carter Niemeyer, who has worked with the species for nearly thirty years and captured the first wolves reintroduced to Yellowstone National Park in 1995 from Canada, also does not consider them to be dangerous at all. In his experience, he has found that 'crawling into their dens, invading their rendezvous sites, catching them in traps where they're agitated, pinning them to the ground when they are fire-eyed and understandably pissed at me ... They've growled some pretty deep, threatening growls at me. But I have yet to experience wolf aggression.'[11]

In recent years, the wolves that have returned to highly developed Western European landscapes have done so without danger to date. It is true that they are few and that this honeymoon may not last but for now no threat to people from their presence has arisen. But they can be inquisitive and the occasional willingness of some wolves to take a closer look at us can be discomforting. Several years ago, I was on a field trip with Sparsholt College in Poland when two of the attending lecturers on a mountain hike one day took some excellent images of an adult wolf walking right behind them on a forest track. They were both agreed that it had posed no threat to them but felt rather that it was simply curious.

On 17 October 2019 the current affairs programme *Transylvania Now* showed a remarkable piece of footage recorded on the mobile phone of Claudiu Luntraru from the village of Csíkmenasá in Romania. In the first of a series of short clips, two large wolves follow Claudiu as he nonchalantly drives his rubber-wheeled horse cart out of the village.

As one runs alongside him, the other comes behind. Both pay keen attention to his large, brindled guard dog, which is chained to the cart. In the final clip, the dog is let loose and bounds across to the wolves. While they close in fast towards it, it's difficult to determine whether their intent is malevolent or not. With raised tails, both sides stand their ground and, when Claudiu dismounts and calls his dog, the wolves melt away. The dog bounds back to him, then returns towards the nearest wolf without any evident degree of alarm. The whole encounter on a sunny day in a landscape of small fields growing arable crops is relaxed and almost playful in atmosphere with no suggestion of menace at all.

There I Saw the Grey Wolf Gaping

Along the Croatian border with Slovenia and Monte-negro, where a population of several thousand wolves still exists, the shepherds and villagers who live among them are also reporting a rise in the incidences of wolves approaching to within metres of people. While some of these encounters have occurred with children who tend livestock, neither they nor the animals in their care have been attacked.

In De Hoge Veluwe National Park in the entirely settled centre of the Netherlands, an individual wolf has recently begun to do the same. Although it's unclear whether this behaviour is a result of illicit feeding, the advances, which began in October 2022, have produced some astounding videos of parents placing push bikes between themselves and the unperturbed wolf, while their kids climb trees in the background. The media attention prompted the national park authorities to propose paintballing any wolves behaving in this manner to both identify those responsible and to scare them away from people. Unfortunately for them, the legal permission they sought to proceed with this strategy was challenged by the nature conservation society, Fauna Protection, who objected to the plan on the grounds of animal welfare and were delighted when its refusal in court meant that, 'The judge chose the wolf and not the humans,' which is really a bit of a historic first.[12]

Be this as it may be, any wolf attack story is still big news. In September 2017 the world's media ran with the headlines that a British headteacher, Celia Hollingworth, who had gone missing after visiting the archaeological site of Mesimvria in northern Greece, had been killed and eaten by wolves.

Celia, who had phoned her family in Britain while her ordeal was ongoing, had said that she had been injured by dogs, but, when the Greek authorities arrived at where she had called from two days later, all that remained was a shattered jumble of bone. Although the photographs taken and shared widely of Celia's remains in a clear plastic sack are truly shocking, the assumption that wolves were to blame was based solely on the statement of a coroner who, never having seen any evidence of a wolf attack before, stated it was his opinion that 'no dog could have administered bites with such force'.

While no DNA was collected at the time to identify either her assailants or any other predators that may have participated in her subsequent consumption, the Callisto Environmental Organization for Wildlife and Nature, in collaboration with the Forest Research Institute, set a series of trail cameras at the site between October and December 2017 to determine the nature of her assailants. The case report they produced for the authorities identified that Celia's location was very close to a seasonal goat farm protected by a pack of livestock guard dogs that contained several very large males. Very little evidence of any wolf presence was identified. As several other previous complaints had been made to the dogs' owners regarding their aggressive behaviour the court, which considered the evidence for her cause of death on 24 September 2022, agreed that the dogs were the most likely predators.

Most of the literature pertinent to the dog breeds that guard livestock throughout the world say little of their ability to harm people. In both Poland and Romania it has been made clear to me by biologists that, even if they are with the sheep and in theory under control, you must be very wary of them. They will bite and, if maintained in numbers, will readily attack people with whom they are unfamiliar. Near Brasov several years ago I watched as a shepherd caught his dogs and put them in an old, green-painted wooden cart on pram wheels before wheeling them through a village in the wake of his flock. All present said that this was normal to prevent them savaging people at random.

A recent study that identifies that dogs kill around 59,000 people worldwide annually – many of whom die as a result of rabies rather than direct attacks – illustrates well their destructive ability. That's 161.64 people a day, far more than lions, which per year only eat around 200 of us, or hippos, which maul a mere 500.[13] Wolves do not even figure on the list. Though the International Wolf Center in Minnesota states that their potential for attacking humans is 'above zero', a recent study undertaken by the Norwegian Institute for Nature Research in 2021 identified only 26 human fatalities – over which half were attributable to rabies – out of 489 recorded attacks over an 18-year period. Most of the incidents occurred in Turkey where twelve people were killed, with another six in Iran, four in India, and one each in Canada, the US, Tajikistan and Kazakhstan.[14]

If you're interested in statistics, that works out at 0.004 people daily.

But there are seven hundred million dogs in the world and only 250,000 wolves. If the tables were to turn, would the latter outcompete the former in terms of gore? Perhaps there is another explanation. Dogs and wolves can and do hybridise widely worldwide. Generally, this occurs when wolf habitats are fragmented and destroyed.[15] In modern Italy, these crossbreeds are called 'bold' wolves as a result of their lack of fear of people and, although no definitive evidence can now be drawn, one famous case of possibly hybrids, or maybe just dogs, killing people was provided by the French Beast of Gévaudan. On 3 July 1764, a young woman named Marie Jeanne Vallet, who was tending cattle in the Mercoire Forest near Langogne in the eastern part of Gévaudan, saw a beast 'like a wolf, yet not a wolf' come at her. The bulls in the herd charged at it and kept it at bay. They then drove it off after it attacked her a second time.

Shortly afterward, its first official victim, 14-year-old Janne Boulet, was recorded and very many well attested attacks and killings followed. Dragoons were despatched by the king to deal with the issue as it rose in significance and, when they failed, professional wolf hunters. On 19 June 1767, a local hunter named Jean Chastel killed a huge red beast with one shot when it broke cover near him during an organised pursuit. Inside its stomach was the broken shoulder bone of a girl who had been killed twenty-four hours earlier. The creature weighed 130 pounds, stood 32 inches at the shoulder, and measured 5 feet 7 inches from nose to tail. The attacks, however, continued until a somewhat smaller female of a similar description was killed nine months later.[16] A study of the records pertinent to the above undertaken in 1987 estimated that 210 attacks been perpetrated by the two, which had resulted in 113 deaths and 49 injured victims. Ninety-eight of those killed had been partially eaten.[17]

It is very possible that chimeras of this sort may for a time have roamed Britain. Do our legends of wolf aggression stem from these?

———

There can be no doubt that in times past wolves would have gorged themselves on the human dead of battle-fields, famines and plague. There is clear testament to this:

When they feed in a country of war... they eat of dead men... and man's flesh is so savoury and so pleasant that when they have taken to man's flesh, they will eat flesh of no other beast.[18]

Perhaps when the conflicts ended, we were 'savoury' enough to become their preferred prey?

On New Year's Day in 1136 there was a battle between the Welsh prince Owain Gwynedd and the Normans on Carn Goch in West Glamorgan where 516 soldiers were killed. In its aftermath, their bodies were dragged around and half eaten by wolves. In the same year, another of Owain's victories at Cardigan resulted in his burning of the town. Its defenders' corpses were said to have been so numerous that they choked the water courses of the marshes through which they had fled and were left 'open to wolves to bury them.'[19] When Owain died, his poets said that the wolves would be among those who mourned his passing as they would have lost their great provider of meat.

Human fodder for wolves would not just have come from battlefields. In a survey of Cardigan made in 1300, the eastern gate of the township was said to have been termed the Wolf's Gate (*Porta Lupus*), and another wolf gate was set into the eastern ramparts of the old city of Chester.[20] English outlaws were called wolves' heads and on occasion, before being executed, had carved wooden wolf masks placed over their faces. 'Let him bear the wolf's head' was a traditional sentence for criminals who could and should be killed on sight, and one fugitive called Godwin was proclaimed a *wulvesheofod* in 1041.[21] Although there are no records at Cardigan of the gate being a gallows, the one at Chester certainly was and, as the suspended corpses of criminals swung and dissolved, nightly visits by the consumers of carrion would have been entirely worthwhile.

There I Saw the Grey Wolf Gaping

But their excellent digging abilities meant that they did not have to wait for flesh to fall. In the Midlands of England, around 500 CE, it was traditional to lay a tree branch over the bodies of the deceased in their graves to prevent wolves or forest dogs digging them up.[22] Many stories tell that wolves excavated graves, and this ability to disinter did not endear them to anyone.

Lying along the coastal tract of north-west Sutherland, where the winds blow in fierce from the Atlantic, is Eddrachillis Bay. It's a lonely rugged spot. Once upon a time, its inhabitants were so troubled by wolves that they had to transfer their dead for burial to the offshore island of Handa.

On Ederachillis' shore
The grey wolf lies in wait,
Woe to the broken door,
Woe to the loosened gate,
And the groping wretch whom sleety fogs
On the trackless moor belate.

The lean and hungry wolf,
With his fangs so sharp and white,
His starveling body pinched
By the frost of a northern night,
And his pitiless eyes that scare the dark
With their green and threatening light.

.   .   .   .

He climeth the guarding dyke,
He leapeth the hurdle bars,

101

He steals the sheep from the pen,
And the fish from the boat-house spars,
And he digs the dead from out the sod,
And gnaws them under the stars.

. . . . . .

Thus, every grave we dug
The hungry wolf uptore,
And every morn the sod
Was strewn with bones and gore;
Our mother-earth had denied us rest
On Ederachillis' shore.[23]

Writing in the late 1700s, the Reverend Alexander
Falconer, a former minister of Eddrachillis, related that
the great stone turrets of the Iron Age brochs that were
erected as coastal refuges from human raiders had also
been used as cemeteries 'from their being a security from
the ravages of wolves'.[24] Similar practices occurred elsewhere in the Highlands and other burial locations exist
on the larger islands of Tanera Mòr in the Summer Isles,
Eilean Munda in Loch Leven, Inishail in Loch Awe and
Inch Maree in Ross-shire. Corpses were no safer further
south. In Perthshire, it was the former custom in Atholl
to bury the dead in coffins made up of five flagstones in
order 'to preserve the corpse from the wolves'. In Assynt,
cairns were built to 'prevent ... numerous wolves from
devouring the bodies' of departed relations'.[25]

In Ireland, there is an ancient tradition still enacted
for funerals at the Gate Cemetery near Ogonnelloe in

County Clare. There, in a walled medieval graveyard located in the centre of a field, the mourners carry the coffin around the perimeter and place it at intervals on the ground to confuse any watching wolves regarding its final place of burial. In Temple Kelly in North Tipperary a similar process also occurs where two coffins, one empty and one full, are used for a similar purpose.[26]

On one occasion, after he had undergone major open-heart surgery, I took Russell Coope for a day out to the West Midlands Safari Park for a distracting wander. He was a commanding figure who liked to bark wayward instructions, which, while amusing, were seldom worthy of obedience.

Although I was aware of this in a sort of distant fashion, as we left his home near Kidderminster his son Robert took me to one side and said, 'Derek, whatever happens today, could you ensure that he brings no more creatures back from the park.'

When I asked why, as I believed it most unlikely that general day visitors would be issued with free tigers or elephants, Robert replied, 'He does it all the bloody time. Had tame badgers when we were kids, one of which attached itself to my brother's foot. After instructing Mum, who was trying to tug the baby vigorously out of the creature's jaws, to lie him on the floor in the hope that the badger might perhaps let go when all movement ceased, he stood over it with a poker to ensure its eventual compliance. Although it did lose interest and the scars on my brother's foot after surgery are today not that bad, I want none of that for my own child. Don't you, whatever you do, let him bring back one more animal to this house.'

We left.

Once out of the disciplined clutches of his family, he assured me while consuming chips and sausages for lunch that this diet was approved by his surgeon and that the cream cakes he had ordered in lavish abundance for dessert were also just fine. We wandered round the park. Russell informed me that the wobblingly fat herd of glistening Nile hippos in a huddle at the edge of their lake were of the same sort whose bones had been unearthed beneath the Ugandan Embassy in London and that these modern elephants from Africa had a very similar tooth structure to the mammoths he had uncovered in gravel pits near Birmingham.

He had boiled their bones in glue to preserve them and believed firmly that an edible soup of sorts could quite probably have been made from this effluvia.

As we wandered around in a contented sort of way, a pal who curated the collection came over. Were we having a good time? Had we seen the new aardvarks? Would we like to see a more unusual creature? A wolf-dog hybrid that had recently been confiscated by the police and which the safari park was caring for pending the identification of a suitable new home.

Russell's eyes lit up.

There I Saw the Grey Wolf Gaping

The beast was bouncingly huge and fluffy. It was pleased to see Russell and he was pleased to see it. It was white with a nose that looked like someone had stuck the bottom of an old man's walking stick to the tip of its very long snout. Gigantic and rubbery. Russell wanted the wolf and the wolf was content. He tried to persuade me that his wife, Beryl, would be fine. I said no. He said that Robert had always wished for a creature of this sort. I said no. In the end he ordered me to transport him and the wolf home. I said no and gave him my mobile phone. I said that if he could obtain a single independent testament from any family member that would support his case then his demand might be possible.

He said no.

We drove home unhappily and I heard no more of it, but Robert said years later that he had remained persistent in intent. In the months that followed Russell took his grandchildren, who generally annoyed him, to see his canine pal. They said no. He tried to take Beryl, but she refused to leave the house.

In the end the safari park staff, tiring of his obsession, threw him out one day on the basis that his rather odd bonding howls were in the end upsetting both the wolf and their members of staff alike.

With that, his dream died.

Lords Do Not Rise at the Crack of Dawn

Kings and great men ... chase fierce bulls,
terrible bears and bold wolves.

Ælfric of Eynsham, *Colloquy*

S arah was an excellent educationalist of the most engaging sort with a viciously biting sense of humour, so when her job with an international wildlife charity in London finished, I employed her to start the schools projects we wished to develop at our wildlife park in Kent.

She loved the wolf cubs, Nadia and Mishka, very much and forgave their every indiscretion. Slippers reduced to rubbery fragments was no problem, pants stolen from her washing basket and shredded was also just fine. When intriguingly odd patterns appeared on her carpets one night after the wolves regurgitated a rancid horse that was also acceptable. She simply opened her windows a bit wider in March to let the gales from the North Sea blow through. The wolves adored Sarah in return and, on nights when all was calm, they would curl around her on the sofa like a Viking queen's robe to all fall asleep together in front of the fire.

Wolves have always hunted wild prey that we considered to be 'our' game and, although at times we cultivated them too for sport, that relationship was always tense. In the Forest Laws of King Canute in 1016, the wolf and fox were specifically identified as 'beasts of the forest or of venery, and therefore whoever kills any of them is out of danger of forfeiture ... nevertheless the killing of them within the limits of the forest is a breach of the royal chase, and therefore the offender shall yield a recompense for the same, though it be easy and gentle.'[1]

In 1066, when William of Normandy left the bodies of Harold and his defeated host on the battlefield of Hastings to be consumed by the 'worms, wolves, birds and dogs', he imposed a new world order.[2] His life passion was hunting and the laws he enacted to protect this pursuit were draconic. No game could be hunted, no wood cut, no collection of fallen timber, no berries or any other forest product could be taken from his preserves. These laws applied not only to wooded areas but also to heaths, moors and wetlands. Transgressors could be heavily fined, jailed, blinded, have their limbs amputated or be put to death. At Stockbridge in Hampshire, a burial dated to the mid-eleventh century contained a man and his hunting dog decapitated together for the illegal pursuit of the king's deer.

William set the New Forest aside in 1079 as a prime hunting land and brought fallow deer with their delicately dappled coats from the warm lands of the Mediterranean to grace the sunlit clearings of his new pleasure grounds. Wild boar and red deer were obtained to join them and when these arrived the wolves were waiting.

In the conqueror's time and for long thereafter, wolf hunts were well-prepared pageants. In the gilt-edged paintings of old hunting books with carunculated trees, tiny and perfect in intricate design; in lightly lined etchings and darkly gouged woodcuts; in faded tapestries woven from wool, the characters play their parts. Dark grey or brindled wolves bound out in front of huntsmen on prancing horses with packs of ordered hounds. Long-eared lymers or scent hounds follow their trail, greyhounds run swift in pursuit, aggressive alaunts with their short, snapping muzzles in many shades straining to pursue as tinkling bells dangle from their neck collars.

For the final confrontation, ponderous savage mastiffs with steel-spiked collars and jangling coats of iron mail.

Footmen follow in blue and crimson hose with lime green hoods. While the riders swing swords, they balance leaf-bladed javelins in their palms. Light and supple, these projectiles were razor sharp. Unbarbed and thrown, they would sever, splice and rupture before falling free easily to enable immediate reuse. Horns ornate with bestial mouths wrought in silver or gold, encrusted and inlaid with jewels were slung on baldric bands around the hunters' shoulders to communicate, direct and blow the '*Mort*' when all was done.

In the end the wolves lie butchered and eviscerated, with crimson flowing. Stuffed with meal, cheese and milk mixed together with their own parboiled internal organs, their cadavers were used as bait for young hounds. Once grown and hardened, the largest of these unleashed against an adult wolf would run straight into or alongside its opponent and aim instinctively for its belly reward with sharp, snapping bites. The kings, lords and bishops who pursued wolves admired their quarry's self-confidence. Adult males in particular would run in a leisurely fashion, often stopping to allow the hounds to catch up in a teasing manner.[3] This façade masked their savagery when they turned in a gully or scrub to rip out the throats of their closest pursuers in an ambush of their own before again moving on. One on one, a wolf would be the match of any hound, but even fights were never a prospect – fresh dogs released in relay after even the boldest of wolves would in the end bring them down.

On many occasions on my visits to the forests of Europe, I have observed the field signs of wolves. The dark shit of their first feed on a kill when they gorge only on the blood-filled organs of the heart, liver and lungs. The bone-, tooth- and hair-filled deposits from the days that follow when they strip bare the bones of their prey. In one of their scats filled with fox fur in the central forest of the Veluwe in the Netherlands, I found a smashed carnassial tooth of its former owner together with the tips of a tiny roe kid's hooves. A dietary jumble of all sorts, crushed, digested and expelled at the crossroads of a track – a territorial warning to incomers that this wolf territory was occupied.

The footprints I saw there differed not at all from the ones I found more recently in Greece in a pine forest awakening from its long winter sleep. Tiny pink crocuses with flaming orange stamens burst their buds through the hard rim of the crystalline snow that bordered the edge of logging roads along which we walked in single file through the mud. Theodore Kominos, a sober, bearded scientist and expert wolf biologist who was our guide for the day, told us that a determining characteristic of wolf tracks was that, while they moved with utter confidence in straight lines, those of the feral dogs that shared the forest rambled and weaved. The fourteenth-century Norman writer Henri de Ferrières, through his fictional character of King Modus, noted that its:

> paws are bigger and rounder than a dogs; the ends of its toes are rounder; and it has a bigger pad and bigger less pointed claws. Its droppings are normally full of fur, because it eats its prey fur and all. The she wolf leaves smaller tracks, but her pad, toes and claws are still broader than a dogs. She leaves her droppings in the middle of a track; the male leaves his to one side. A young wolf has tracks like the she

wolfs, except that the claws are longer and more pointed.[4]

These field signs accurately described would have been familiar to William of Normandy.

He would have observed the trails that led to their forest lairs. Beaten down, weaving through tall vegetation. Passing over fallen trees, around rocks and boulders to finish at the edge of a carrion-strewn rim of soil above a dark void in the earth. Perhaps he witnessed his huntsmen going down into their dens to impale shrieking whelps on the double-pronged gaff, still employed in Brittany for this purpose less than a century ago. Once rammed through a ribcage its broad blades could only be freed by dismembering.

He would have known of the other common strategies employed in their pursuit. To lure a pack into an ambush, his hunt masters would have provided a dead horse as bait. After hauling its cadaver to the required location, its legs would be cut off and then slowly dragged to all four sides of the forest before being returned once more to lie alongside the central torso to establish an alluring scent trail. Once a wolf pack's whereabouts was established, this same process would be repeated for several nights to persuade them to stay and enjoy their equine repast. They were cunning, however, and did not always choose to do so if they suspected a trap. When disturbed, as one hunter plaintively observed, they might 'eat in the night and in the day they will go a great way thence, two miles or more, especially if they have been aggrieved in that place, or if they feel that men have made any train with flesh for to hunt at them.'[5]

In the evening before the hunt, what remained of the horse would be hung from a tree by a rope to make any last meal difficult to obtain while the larger bones placed on the ground underneath it maintained the interest of the wolves. In the early morning the hunt master would creep back and cut the carcase down in an effort to guarantee that, prompted by hunger, the wolves would rapidly return. As Gaston Phoebus III, the thirteenth-century count of Foix who was a hunting fanatic, observed, lords commonly 'do not rise at the crack of dawn' and the wolves might therefore need to be retained in the wood for his eventual pleasure by lighting fires around the perimeter.

When all was correct and the mighty were ready, the first leash of greyhounds would be positioned down-wind before the hunters entered the forest with yelping hounds to drive the wolves on. Horns were blown to speed their flight and encourage the dogs. When the wolves broke cover, the hounds held in a diamond pattern were released in relays, some once the wolves passed them, some when they were level. The last were released directly in front of their flight line to confound any prospect of escape and to bolt them in confusion into a system of pre-erected nets. Watchers waiting there with two sticks would throw a straight one in the wolf's face before using a forked one, once they were well entangled, to pin them down by their necks to make stabbing straightforward.

Wolf hunting was dangerous and William's son Henry I (1068–1135) stipulated that: 'If a person takes a bow or a spear from a hunter, or sets a foot-trap against wolves or something else to be caught, and receives any damage or evil, he shall get payment for what happened.'[6] During the reign of his great-grandson Henry II (1154–1189), awards were still being made to the hunters of wolves in the New Forest.

While the laws made by William endured, wolves were protected, but when in time their onerous restrictions were repealed everything changed. Without penalties, the abundance of deer, boar and other prey plummeted and those who still wished to hunt had to develop a new approach. Rather than pursuing game freely through whole landscapes, parks with stone or earth-banked walls and palisades were erected to harbour and protect the prime quarry species. Though uncommon in pre-Norman England, the Domesday Book of 1086 refers to two different types of game enclosures: parks with fixed and complete barriers and 'hayes' of a more temporary nature along which game could be driven to hunters in ambush. While Domesday records over seventy 'hayes', it mentions only thirty-one parks. By 1300, there were more than three thousand and, although William Harrison writing in *Holinshed's Chronicles* in 1577 complained 'that what store of grounde is employed upon that vayne commoditie which bringeth no manner of gaine or profit to the owner', he was ignored and emparking continued.[7]

Although medieval parks covered huge expanses of land, the deer and any other beasts they contained were not free. These were property and as such jealously guarded. Legions of keepers and wardens were employed to ensure their security and deny them to people and predators alike. When Edric de Wholseley occupied Wolseley Manor in the eleventh century, it was granted to him as a reward for 'killing all the wolves which were ruining the King's hunting by preying on the deer in the County'.[8] In 1263, the Sheriff of Stirling employed in repairing and extending the Royal Park in his borough engaged a wolf hunter to ensure the safety of the deer. Alane de Wulfhunte held land for the service of destroying wolves in Sherwood Forest at the behest of Edward III (1312–1377) and in 1433 Sir Robert Plumpton held a bovate of land called 'Wolf hunt land' in Nottingham, 'by service of winding a horn, and chasing or frightening the wolves' in Sherwood Forest.[9]

During the reign of King James I of Scotland, an Act was passed in 1427 for the destruction of wolves in the kingdom to protect in part the greater game enclosed in the parks such as those at Falkland and Kincardine.

While the emphasis of these records focused on game protection, other social pressures began to arise. The refuge that forests offered to outlaws and criminals was becoming a serious consideration. In 1439, Robert Umfraville held the manor of Otterburn on the basis that he ensured Riddlesdale was free of robbers and wolves.[10] The monks of Coupar Angus Abbey, in a lease of their lands of Innerarity in 1483, required tenants to 'obey the officers rising in the defences of the country to wolf, thief, and sorners'[11] as the crimes of all the foregoing were considered to be common.

In around 1427–8, a seemingly vigorous act of the Scottish Parliament, which extorted that 'the baron within his barony ... seek the whelps of the wolves and slay them ... and when the baron ordains to hunt and chase the wolfs the tenantry shall rise', is somewhat softened in the intent by its last sentence, 'and that no man seek the wolves with shot but only in the timings of hunting of them.'[12] Did a tantalising yearning to retain some remain?

The French, who ritualised wolf hunting as a noble sport, also faced the same contradiction. To kill but not so much that your desirable target, however tiresome its presence might be for others, disappeared. In the time of Charlemagne in 813 a body called the *'luparii'* was created to act as agents for the management of animals that damaged his imperial estates. Later in time, this morphed into a sort of field gendarmerie called the *'lieutenants de louveterie'* or 'wolf catchers'. Well into the 1900s, its members, who were typically drawn from the aristocratic hunting classes, mixed with ex-military sorts who sported grand moustaches and green uniforms with shining brass buttons.

In the aftermath of the Revolution, drives organised by those of them who kept their heads were augmented by British fox hunters who crossed the Channel to swell their seasonal ranks. Although the theoretical role of the *louveterie* was to destroy wolves using all means possible, certain methods of killing such as the use of poison and sprung steel bow traps were viewed with disdain. Even in the districts where their hand was hard there was a common saying, *'En France, nous "préservons" nos loups, pour le plaisir de les chasser à courre,'* or, 'In France we "preserve" our wolves so as to be able to enjoy hunting them.'[13]

But if you are just beginning to imagine that persecution of this sort might have produced a clear trajectory of decline in wolf numbers, think again. Hunters of all sorts have always attempted to augment the object of their pursuit. That they do so today with sad farm-reared pheasants and pitiful green mallard ducks that wheel and turn back to the waters they have flown from again and again to face a succession of blistering fusillades is both obvious and nauseating. It is more than likely in Britain that they did so with wolves that, once captured, could be recycled. In 1167, John Cumin was paid ten shillings by the bishop of Hereford for the capture of three wolves and three years later Norman the keeper of the Veltrar hounds and William Doggett received a similar sum for the capture of two in the forest of Irwell. While the fate of these captives is not clear, there could have been a distinction between the killing of wolves and their capture alive, as Phoebus makes it clear that by 'such diverse engines' the living might afford further sport still when released into game parks.[14] That their manipulation was undertaken for just this purpose was verified in 1465 when an individual called Bohemian Schaschek writing about a southern English location stated that 'this country will not support wolves. If they are introduced, they die forthwith, which has been proved by custom and experiment.'[15]

Even later in time when home-grown wolves were no longer available, they could be imported from abroad and purchased from dealers in wild animals. It is very likely that Colonel Thornton (1757–1823), who owned a sporting property called Falconers Lodge in the Yorkshire Wolds into which he proposed to import a breeding stock of wolves to release, knew this. When 'his farming neighbours responded with a furious and bitter hostility' to his proposal, he paid little heed to their concerns.[16] While his hunting diaries cryptically recorded that wolves might only be pursued along with roe deer on Tuesdays, the *'brace of wolves which formed the advance guard'* made a truly memorable impression on the spectators of his French hunting extravaganza in 1802.[17]

Other much odder things followed. Several landowners were involved in crossing wolves and dogs. In a garden corner of Wilton House in Wiltshire is a gravestone:

Here lies Lupa, whose grandmother was a wolf,
whose father and grandfather were dogs,
and whose mother was half wolf and half dog.
She died on the 16th of October 1782, aged 12 years.

Lupa was born in 1770 and was given to the Earl of Pembroke. At Wilton, she had her own keeper, Nathaniel Townsend, who was paid between £1 and £2 every few months for 'the keeping of the wolf-bitch'. Although she was kept separately from the other hounds to avoid any fights, it is unlikely that she lived in the house.

Lupa reportedly had four litters of puppies, one of which is recorded in the household accounts in 1773. Lupa was the result of an experiment performed by Joshua Brookes, a menagerie owner and animal dealer in London who owned a male wolf and reportedly bred it with a Pomeranian bitch. This coupling resulted in nine puppies, another of which was acquired by Lord Gordon and taken to Gordon Castle in Scotland, where it was seen by the naturalist Thomas Pennant. He recorded that it 'resembled a wolf and being slipped at a weak deer instantly caught the animals throat and killed it ... Another of the same kind ... stocked the neighbourhood of Fochaber with a multitude of curs of a most wolfish aspect.'[18]

Edward Lewis Davies was the sort of hunting parson who sailed annually to Brittany in the mid 1800s to hunt wolves and wild boar. One night at dinner, he enquired of the Count de Kergorlay about the breeding of some particularly large savage hounds he had hunted with that day.

'A big, bold, broken-haired hound is what I keep for the work; and occasionally I invigorate the race' said he, 'with a strain of wolf-blood.'

'And how, pray,' I inquired, 'do you manage that?'

'Nothing is more simple ... I keep a dog-wolf brought up by hand; and he, suckled in infancy by a hound dam, lives in perfect concord with any hounds I think fit to enclose with him in his kennel.'

'And do you find the first cross,' I asked, 'as manageable in chase as your ordinary hounds?'

'Far from it,' he replied; 'insomuch that I only keep that produce to breed from. They usually run mute ... and are so self-willed in chase and so fierce in kennel, that I merely use them as stud hounds, and enter the second cross.'[19]

Davies then stated that a Mr Waldron Hill brought over several couples of these hybrid hounds from the French Department of Eure, which 'had at least one other fault, which he found utterly ineradicable ... They *would* kill sheep; and, as he justly remarked, that to hunt the sheep was a far more expensive amusement than hunting the otter, he hanged the whole of them.'[20]

In 1861 it was reported by the newspapers that a young wolf had been caught at High Ongar, in Essex, and that others had been seen thereafter in the fox coverts of the neighbourhood. The first individual, which was preserved in Chelmsford Museum as a 'young wolf killed in the woods near Ongar after committing several depredations', can no longer be found. One explanation at the time for its presence was that some wolf cubs were included among the foxes imported from abroad. This may not have been entirely accidental as the import of wolves that were then released for hunting purposes 'has often happened' according to a Mr J.C. Bellamy'. In 1862 the plot thickened when another creature captured by a carter in Ongar Woods in the presence of two others grew into 'an undoubted Coyote' or 'in reality a north African jackal'.[21]

Abraham Dee Bartlett (1812–1897) in his long coat and top hat was the superintendent of London Zoo for many years. His authority as a natural historian was based on the vast experience he gained directly from the observation in life and dissection in death of a wealth of wild animals from every corner of the globe. After examining the carcase of one of the Ongar enigmas Bartlett found it somewhat disappointingly to be 'a common fox, without a trace of any other animal'. Although you would have assumed that he was not the kind of individual to get his prairie wolves, jackals and wolves confused, the good lithographic illustration of one of the creatures sitting erect with a bone protruding from its mouth, its front legs outstretched and long bushy tail curled under its rump in *The Mammals, Reptiles and Fishes of Essex* written by Henry Laver in 1898 looks more like a jackal than a fox so perhaps this mystery endures.

The *lieutenants de louveterie* survive – even though the last wolf in Brittany died at Carhaix in 1912, there are currently nine still employed in Finistère. They have become volunteers working for the state and now are charged with regulating harmful animals and maintaining the wildlife equilibrium. While a report of a female wolf near Lac de Guerledan in the Morbihan region of Brittany[22] in 2018 was never confirmed, in April 2020 a single wolf was photographed on a trail camera strolling through the village of Londinières near the Channel coast, and on 3 May 2022 another was filmed on a camera trap in the commune of Berrien in the Arrée Mountains.[23]

Their long trek home from extinction has begun.

One day soon, they will breed in Brittany and the hobbyists who keep flocks of tiny nut-brown sheep from the Île d'Ouessant will have to lock them away for the night.

Will the *louveterie* meet these wolves with bludgeons? Do they already have pitch bubbling black at the ready?

Or, at the dawn of the twenty-first century, at a time when our relationship with the wolf is changing, will they greet their old enemy with a casual '*ca va*'?

CHAPTER SIX

The Lord Is My Shepherd

The sward the blackface browses,
The stapler and the bale,
The grey Cistercian houses
That pack the wool for sale.

John Meade Falkner, 'Cistercians'

I don't dislike sheep and, although I no longer keep anything like the number I used to, a small flock of Shetlands in piebalds, blacks and browns is grazing outside the window as I write. Their summer job is to mow the lawn and, as an aside, provide an annual crop of rather tasty lambs.

I do not participate in the annual purge of foxes that takes place on the farms surrounding mine as they are the catchers of the voles that debark our young trees and as such are very welcome. I don't begrudge them an odd lamb if they can summon the will to steal one from their foot-stamping, aggressively butting mothers and so my sheep are fast becoming part of nature.

From the earliest of times, farmers have always fought to protect their livestock from predators. On Dartmoor in a landscape little changed from the Bronze Age, the great boundary wall of Grimspound still stands. Over five hundred feet in length, ten feet thick and when new up to five feet in height, it was faced with large slabs laid in horizontal courses, with a core of smaller stones in between.

The settlement it protects, with its twenty-four houses, overlooks a valley to the north where grazing was plentiful. The stream running through its northern part would have watered people and their animals alike when they withdrew for the night. Though Grimspound encloses four acres, Broadun nearby contains more than fourteen and the remains of over a hundred others of their sort are visible to this day. Despite their appearance they were too weak to fortify against any attacks by a human enemy. The walls were instead erected to protect from a common foe. Alert, cunning, swift and determined, the wolf was the herdsman's adversary.

In June 2020 I went with Stephen Trotter of the Cumbria Wildlife Trust to look at a wildflower meadow restoration project at Eycott Hill near Penrith. That day, the delicate heads of meadow cranesbills with their long beak-like stamens nodded in cobalt splendour in the soft gusts blowing warm from the fells, while melancholy thistles with tight punk-purple headgear bowed back.

When we were done admiring these marvels, we walked over the rock ridge that bisects the reserve to view a wetland below that might one day, Stephen hoped, host reintroduced water voles.

As we paused on our path, I noted without thought the scattered base rocks of an ancient corral. Lichened grey with age and unmarked on our map, its single stone entrance led into a circle of nearly fifty metres in diameter. We walked its bounds and, when near halfway, we found to our great delight what we thought we might not. A rectangular chamber of approximately three metres by two joined into its inner side. A shelter too small for a flock of sheep or the small dark cattle that would have been driven inside before nightfall.

This was the lair of the watchers. Forgotten by us but remembered far to the north in the Hebrides where the people once knew of a time long ago 'when they had to kindle ... the wolf-scaring faggot that guarded.'[1] Fire was ever our friend in the dark, giving light, warmth and a weapon of force if required. It's been our crutch from the earliest of cave times and we have used it for centuries to flare beacons of warning, guide travellers to sanctuaries, illuminate crossroads and light the outside of our door-ways before opening them at night to the dark.

Though no memory remains of the watchers at Eycott, they would have realised its worth.

With a layer of bracken on the floor and a roof of heather turf, robes laid inside would have offered relief from the hard ground below. If a small fire was possible, heat or the hope of a warm meal for both master and dog. In the good times of plenty during the short nights of summer with a sound wicker gate, all would have been calm. Sound sleep for those inside. Slumber untroubled for the assistants at the gate.

But when the long nights of winter met the searing cold of a slow spring's growth, vigilance was all. Howls in the distance were not to be feared. They were the norm, offering no portent of imminent attack. The worst was the silence. Broken when the dogs with their nail-studded collars raised their heads from slumber to snarl and run. When torches thrust into the embers burst forth into blaze. When spear shaft, bow or leather dagger hilt was grasped firm to face the foe from beyond the walls. You had to be quick. The dogs snarling and leaping against the internal wall could smell what was outside. The sheep clustered tight in reeking fear. Perhaps they would come from one direction. Perhaps from both sides. Companions might run to your aid. If you were alone then the battle was yours.

Terrifying, swift and ruinous if the wolves scaled the walls or dug under the branch piles that infilled the gate.

If your torches flared fast you could see.

If they failed you could not and they were inside.

They would test you. They were clever and strong.

But no readied attack tonight. Perhaps tomorrow or in the dark of another night when the surprise was with them.

Breathless it would be over.

Your fold would pantingly calm. While the sheep in their stupidity would settle to sleep and the dogs might cautiously curl, you would know no ease until first light when the flock would flow out once again to the high tops.

———

When the white monks walked to Britain in 1128, it was to found their first British settlement at Waverley in Surrey. The next, begun on 9 May 1131, was in the prime sheep country of Tintern in the Welsh borders. The Cistercians took the name of their order from the Latin word *Cistercenses*, derived from the name of the town Citeaux in southern France where their community began. Unhappy with a perceived laxity in the observance of the rules of St Benedict, they abandoned the frivolous ways of his followers to commence a much stricter regime of their own. So pitiless was the austerity they embraced on their *'surest road to Heaven'* that they would only don robes of no colour.[2]

They were sheep farmers and loved anything that disturbed the placid sanctity of their flocks not at all. The wool from the vast church flocks was their gold, as Susan Rose discusses in her book on the medieval wool trade, *The Wealth of England*. English wool, although not all of good quality, was nevertheless available in sufficient quantity at a median standard to ensure that it outcompeted any other supply available in Europe. Some of the highest worth came from the flocks surrounding Abbey Dore. The wool trade had a huge impact upon the economy and society of medieval Britain. At its golden peak between the mid-twelfth and thirteenth centuries, wool production formed the nation's financial spine. The combined flocks of the church lands were vast. Fountains Abbey in North Yorkshire had no fewer than nineteen thousand sheep at one point in the fourteenth century, while its neighbours at Jervaulx, Rievaulx and Byland kept in the region of thirteen thousand each. The monks' reputation for taking great care of the wool they produced to ensure it was well washed, dried and sorted according to quality meant that they could command the highest of prices. While the trade flourished, it funded not only the abbeys themselves but ancillary churches, castles and estates.

Although it's hard to appreciate now if you keep sheep in a time when even the finest of wool is worth much less than the payments for those who shear them, wool was for a time the wealth of our nation. The once pre-eminence of sheep that ensures that even today the presiding officer of the House of Lords sits on the Woolsack began a near millennium-long love affair that has shaped quite utterly both our landscape and minds. Everywhere it could be, wool was produced for export to the centres of cloth production in the Low Countries, France and Italy.

Shepherds were carefully chosen and goodly tomes of well-considered advice were made readily available to their employers. They must not leave the flock 'to go to fairs, markets or wrestling matches or to spend the evenings with friends or to go to the tavern'. If flock owners noticed that sheep were shunning the shepherds it was 'no sign that he is gentle unto them'. Sheep were delicate creatures much prone to a broad range of diseases from coughing fits to internal parasites, scab and lice. Their progeny could be eaten by 'ravens, kites and crows' if they were born in a field rather than a fold. Plants such as gorse would be 'very harmful to ewes in their throat's gullet' and poppies 'with a round leaf and red hairy stem made them ill in June'.[3] They had to be guarded against other dangers as well, such as becoming 'too hot or too damp', which would entail moving them to high pastures or cooler areas of shade.

Some churchmen formed relationships with wolves. When St Botolph or 'Botwulf of Thorney' retired to lead a hermetical existence in the great marshlands that covered Suffolk in 600 CE a large wolf assisted his focus when marsh daemons tried to distract him from his concentrations on God. This beast became his soul mate and Botolph as a result is one of a very few saints to be depicted in early iconography as having a wolf's head. He was the founder of the now derelict abbey at Colchester and would have been proud to know that in modern times he has become the logo of its railway station.[4]

St Francis of Assisi also preached kindness to animals and willingly travelled to Gubbio in central Italy to meet a wolf that had been terrifying the townsfolk and preying on their livestock. After listening carefully to the villagers, the saint sauntered out to meet the wolf, which those left behind were convinced would consume him. St Francis, however, had God on his side and began a reasonable discourse, which established that, as the wolf was too old to hunt wild animals and just needed to eat, it was more than willing to cease plundering if the inhabitants of Gubbio provided it daily with a decent square meal. Francis agreed to its terms and, after informing the incredulous townsfolk that 'God is the wolf's God too', departed to undertake more good deeds elsewhere. In the lower area of the modern town is the church of St Francis, in the piazza of which stands a bronze of the saint stroking the wolf that stands by his side. The first church, which was built in the second half of the thirteenth century over the house of a family that welcomed him when he left Assisi, was excavated in 1873. During the dig, a large canine skeleton believed to be a wolf was unearthed. At the local community's insistence, it was reinterred at the back of the church, where its tomb remains visible to this day.[5]

Although St Francis's wolf directive worked for him elsewhere, other churchmen with more earthly interests were beginning to ask God what should be done about wolves. His advice was that: 'No life can be pleasing to God which is not useful to man.'[6] Jesus made his position clear by confirming that:

I am the good shepherd; the good shepherd lays down his life for the sheep. The hired hand is not the shepherd, and the sheep are not his own. When he sees the wolf coming, he abandons the sheep and runs away. Then the wolf pounces on them and scatters the flock.[7]

The early iconography of Christianity so well developed the pastoral symbolism of sheep as being both valuable and vulnerable that it remains, despite any logic, nearly utterly unquestioned to this day.

In the library of Canterbury Cathedral, an illustration of a fox running off with a chicken pursued by a medieval peasant with a club is twinned with another of a wolf running off with a ram slung over its shoulder, its shepherd departing swiftly in the opposite direction. Other illuminations depict them creeping up on folds while guard dogs and shepherds slumber. Large flock owners employed nightwatchmen with bows and hounds and it was these individuals rather than the daytime shepherds who were required to be extra vigilant at dawn when the grey hour that the French describe still as being between 'dog and wolf' came round.[8]

Wolves were outside the world of order that the Christians required. But what to do when the production of livestock was disrupted and generous rivers of resultant income were confounded? Though prayers beseeched God, these were of little practical worth.

Lift our flocks to the hills,
Quell the wolf and the fox,
Ward from us spectre, giant, fury,
And oppression.[9]

In a time when the medieval church controlled nearly every aspect of one's life, few would have questioned this view. Furthermore, church forests that produced valuable timber offered wolves easy refuge and, most gallingly if you were one of the hunting clergy, your dainty deer might be savagely abused. This last insult was insufferable and when St Hubert, after a hard night in his distillery, encountered a stag with a glowing cross between its antlers the game was up.

Kill them all.

The enclosures where sheep spent the night were called cottes or folds. In Abernethy Forest in the Cairngorms, Scotland, the naturalist Roy Dennis has seen and photographed a number of these structures sitting high on raised mounds. Three generations of the farming family who graze sheep there still call them the 'wolf cottes'. In England similar structures in high Cravendale, where wolves were killed as late as 1306 and existed without doubt for very much longer, also stand firm.

But more remarkable than these by far is the wolf wall. Located at Lower Winskill Farm in the Yorkshire Dales, surrounding a complex of Cistercian sheep pens, is a barrier made of rocks nearing 1.6 metres in height. Its base is over half a metre in width and it stands on a series of large upended stone slabs called 'orthostats'. Nothing can dig under it, and at its top wide, flattened flagstones projecting outwards for over eight inches ensure that nothing coming from the outside can climb or leap over. Built in the early 1300s, it is believed to have been maintained for nearly three hundred years before falling out of use by the beginning of the 1600s. Perhaps, like the folds in modern Romania where the wolf threat to sheep in the mountain pastures is ever present, a rampart of turf into which are set sharpened stakes may have provided its structure with an additional assurance of security.

Bells as an age-old mechanism for registering the movements and activities of flocks or herds are largely forgotten in efficacy as a form of communication. Years ago, in Portugal on a sunny mountain terrace outside a delightful bar I sat drinking cold beer with a commercial carrot farmer. Blue-spotted geckos skimmed the white walls of our lookout with its views of the deep lower valleys flanked by scrubland on their sides and crests. Griffon vultures turned lazily high in the turquoise sky. Both the languid creamy cattle and the tight knit flocks of multicoloured sheep, which moved rapidly heads-down in dry, dusty mobs, wore bells.

While the sheep tinkled in light jingling carnival, the big clappers in the cowbells clunked slow.

'That's the way you want them,' said the farmer. 'You can hear the rhythm several valleys away. When it's slow all is fine. When it's fast and confused there is a problem.

When you hear no sound at all and there are vultures overhead then you know you're too late.'

In the flower-rich mountain pastures of Europe, the age-old practice of transhumance – the driving of live-stock from the lowlands into the mountains for summer grazing – leads modern herders quite commonly into seasonal conflicts with wolves. Transhumance was once common in Britain and Ælfric›s tenth-century collo-quy clarified that it was the shepherd's lot to 'drive my sheep to their pasture, and in the heat and in cold, stand over them with dogs, lest wolves devour them'.[10]

While now we assume that those who watch sheep to ward against the dhole or the jackal, the hyena or coyote must hardly like them at all, this too is not clear. There are exceptions bound by ancient understanding. In the mountains of Carpathian Romania many of the older shepherds accept without rancour the annual loss of a few sheep as their due to the wolves. In their view the wolves have been there forever and always will be. Maybe in the beginning we too were prosaic.

———

In the late 1500s, John Caius was advising in his book *Of Englishe Dogges* that while elsewhere 'the shepherds hounde is very necessarye and profitable for the auoy-ding of harmes and inconueniences which may come to men by the means of beastes', in England at least 'Our shepherdes dogge is ... of an indifferent stature and growth, because it hath not to deale with the bloudth-yrsty wolf'.[11] His thinking was wishful, and practical sheepmen would have been well advised to dismiss his assurance and pay heed to the Roman agronomist Columella, who in *De Re Rustica* advised shepherds to choose white or light-coloured breeds for maximum visibility when fights in the dark were the norm.

In the velveteen hills of the Lammermoors in the green Scottish borders a book of ancient traditions instructs still on how to train these flock dogs. Taken early from its mother, a selected puppy would be suckled on a ewe several times a day. Nests would be made for it out of sheep's wool and the puppy would thereafter be allowed no association with either other dogs or people. Inte-grated into the flock, the sheep, their scents and their daily rhythms would form the orbit of the young dog's life. When it grew to an age where meat became part of its diet, it might only be fed from the pens that the sheep were brought into every evening. Once established on this course, the dog would defend the sheep ever after from attackers of any sort.[12]

Sheep dominate now, as they have done since the time of the Cistercians, the land use of much of upland Britain where a 'good shepherd' mindset remains firmly in vogue. This myth of benevolence masks a much darker history. Thomas More in his book *Utopia* described how the sheep's 'raging appetite' for the common pastures of the poor folk had turned them 'into man eaters. They destroy and devour entire fields, houses and cities'.[13]

In his Tudor time, the desire for the financial rewards of wool led landlords to dispossess whole communities of tenants to accommodate sheep and large numbers of rural people were forced as a result into destitution, criminality or a pitiful starveling's death.

The national sheep flock in Britain is vast. Around twenty million breeding individuals, it rises by another ten million annually when their lambs are produced. The Dutch sheep flock numbers in the region of five hundred thousand and is believed to be falling. In the Netherlands, where on average thirteen thousand sheep are killed annually by dogs, Dutch farmers were reported in 2018 as becoming 'alarmed' when wolves killed 138. On 23 January 2017, *Farming UK* magazine reported that approximately fifteen thousand sheep had been killed in 2016 in Britain by dogs and, as most are never bred to live long lives in any case, presumably around ten million by us in England alone.[14] If the rationale in Britain was to rid the landscape of wolves to ensure that sheep would forever graze unmolested it has singularly failed.

In May 2018, under the title 'Killer Sheep Discovered Wolfing Down Bird Chicks', the popular BBC television series *Springwatch* produced a short film of sheep trampling or directly consuming the chicks and eggs of both curlews and lapwings. Series presenter Chris Packham also spoke of another study that found sheep on the Shetland island of Foula killing and eating the chicks of Arctic terns.[15]

In truth there is nothing to recommend the overwhelming contemporary presence of huge numbers of sheep in the British landscape. Their trampling of uplands results in soil and scree erosion and destroys the formation of deep moss communities that hold and absorb the flow of water. The massive silt run-off from the fields of root crops used to fatten them in the winter months pollutes rivers and smothers the spawning beds of pearl mussels and migratory fish. Sheep consume tree seedlings with consummate ease and, when more hardy thorns attempt to grow in response, their keepers burn these to ash. Every pond, pool or wet hollow there could be in pastures is drained to eliminate the mud snails that sheep eat inadvertently when the grass is short, in an effort to ensure that the cysts of the liver fluke the snails carry cannot infect them. Add to this the regular dosing of sheep with cocktails of toxic chemicals to kill the insects that parasitise both their skins and intestines and it can be fairly stated that the system of sheep pro- duction we practise in modern Britain is nothing short of catastrophic.

And for what.

Well, while at one time their wool was valuable this no longer is the case. Commonly its value is so little that it does not cover the shearing costs and any transport for sale is pointless. Increasingly, as a result, wool is thrown away, buried or burnt. Although fertility in the past was so effectively derived from sheep dung in nutrient-poor soils that they were called the 'golden hoof', this was in a time before chemicals and is no longer a driver of most farming systems.

Their meat is increasingly undesired and, at around 10 per cent of our red meat diet in any case, we no longer consume much of it.[16] Without an export trade and eye-watering levels of farm subsidies drawn directly from you, the taxpayer, sheep farming, despite its bucolic image of frisking lambs and dogs being whistled at in sunny spring meadows by ruddy-faced chaps in flat caps, would simply not survive in its current form.

Although farmers are commonly coy about revealing the figures they receive from the public purse, at the time of writing, I get approximately £17,000 annually under the old system of land-based payments for doing nothing much and around £34,000 for turning my meadows into insect-rich havens for wild flowers and birds, which is an infinitely more joyous pursuit.

My cubs grew into happy wolves. Mishka was always pleasant but more distant than Nadia, who was forever full on. Unconvinced that you were ever clean enough, she would stand on her hind legs, place a paw on each of your shoulders and lean down with her great head to ensure that your countenance was, by the time she finished licking it, as pristine as possible. When we could we fed them entire carcasses to ensure that their digestive systems had bone and hair flushing through. One day we received a large Suffolk ram that had died on a nearby farm. This feast of some size, which would last several days, took two of us to lift its large black-legged, bulging body into the wheelbarrow to push it up to their enclosure. When they saw the delight coming, both wolves ran alongside us, tails up, along the perimeter fence. We opened their double doors, shut the first behind us, pushed the ram inside and then tipped it out of the barrow. After greeting us in a somewhat perfunctory manner Nadia, who was hungry, simply picked it up by the back of its head in her jaws, lifted her head and then walked off with it suspended, feet flopping, into the forest to consume it in peace, while Mishka followed behind.

CHAPTER SEVEN

Most Likely a Dog

This black dog, or the devil in such a likeness.

Abraham Fleming, *A Straunge and Terrible Wunder*

On 21 February 2015, Ben Coult of the Moldy-warps Speleological Group and Paul Rodrigues of the Durham Cave and Mine Club were exploring a location called Swindale Pots to the north of Brough in Yorkshire. In the vicinity of an area called West Sink, they observed a narrow fissure among some boulder rubble in a low, well-vegetated gully. After moving a few large rocks, they passed down into a small chamber.

In the light of their torches, among fallen rock and detritus, lay a dirt-covered skeleton. Its skull was large and dog-like with long gleaming incisors. As Ben was interested in natural history and knew that a team from York Museum excavating potholes and caves nearby had identified the remains of bears and lynx, his first thought was that this was a wolf. In a search for further evidence, they squeezed in to a point where fallen boulders almost closed the chamber's space at its end. Slithering into a gap beyond these, they could see in the shadows, just out of reach, an additional skull of a similar shape along with associated bones at the end of what looked like a burrow in a slippage of limestone.

As they had no tools and the hole was nowhere near large enough to access, they withdrew.

The following weekend they were back. Accessing the second chamber with tools was easier than they expected and, after photographing the skull and skeleton in situ, they bagged the remains. Ben took the skulls home and cleaned them before sending pictures to a number of archaeologists who were positive that they were wolves. A pal of Ben's called Alex, who was studying for his PhD in carnivore taxonomy at Durham University, had excavated modern wolf lairs in North America and confirmed this view, suggesting that from their skull size they were probably young adults.

Further excavations in the dirt floor of the cave uncovered visibly gnawed bone fragments from sheep, deer and a small cow. Both of the skulls were missing their brain cases and, on enquiring of caving experts, Ben was informed by one, Tom Lord, 'that as both skulls had their vaults smashed it was possible that the shake hole had been a lair site … utilised as a wolf trap in medieval times.'[1] That the natural circular pits that occur in the limestone karst may have been used as wolf traps in the past is suggested by a place name of Ulphpits – Ulph is old Norse for wolf – nearby. Although no definitive cause of death could be identified, the skull at the front of the cave had a canine socket that was not big enough for its tooth. In time, this deformity had resulted in a wider abscess that had distorted its front teeth. While this may not have caused its demise, while it lived it must have done so with significant pain.

I first came across Ben's story in April 2020 and had no idea at that time what a remarkable find it was. We got in touch and his father, Terry, who is a naturalist of standing, informed me that, while Ben still had the skulls in his bedroom wardrobe, they had done no DNA or radiocarbon analysis in case they were simply the bones of large dogs dumped by farmers, who often use open shake holes for rubbish. After discussing for a bit, we agreed that it seemed incredibly unlikely that even the keenest and slimmest of dumping farmers were going to drag two big dead dogs that far underground and that it was much more likely they had made their own way in.

I decided to seek expert assistance and contacted an old colleague from times past, Dr Danielle Schreve, who is the current Professor of Quaternary Science in the Department of Geography at Royal Holloway University of London. Amazingly well informed on all matters pertinent to the history of mammals in Britain, Danielle is also a jolly sort. Adept at drinking malt whiskies and telling funny jokes, she had also been a colleague of Russell Coope and used to visit him when he retired to Loch Tummel in Scotland.

The first time I stayed with Russell and Beryl, he told me he had every native species of British land mammal on his property. Having warned me that the water coming out of the bath taps was very likely to be orange in colour but not to worry because it came straight off the Perthshire Hills, he then proceeded to tell me that I should also be alert for tiny skeletons. This prospect was slightly more alarming until he explained that bats regularly fell into the water tank. I confess I waited in trepidation to see what emerged but as far as I know, I was the only mammal in the bathtub that night.[2]

Danielle wanted to see both the photographs of the skeletons as they had been found and then the remains themselves. Her initial view was that, as they were simply lying on the surface, they had died there naturally. She was clear that the absence of a brain case in both did not mean that they had been killed by people as she had seen many illustrations of Pleistocene wolf skulls with the same type of entirely natural breakage.

But wolf and dog remains are not always easy to tell apart, and while many older antiquarians once accepted any large canid skull found in a remote location as a wolf, we now know as a result of DNA analysis that many quite simply were dogs. The different forms, shapes and sizes of dog breeds complicate matters.

Though the surrounding context of the location of any uncovered remains is important, even this is not always as simple as it seems. In a boundary ditch dated to a period between the sixteenth and eighteenth centuries, the remains of a large dog were unearthed near Witcham in Cambridgeshire. The bones were believed to have belonged to a male dog, as suggested by the exaggerated sagittal crest or bony ridge running along the middle of its skull. While detailed physical analysis in 2008 implied that it was nearer in type to a wolf than any other breed of dog known at the time, it has not yet been tested for DNA and as such this assumption is unclear.[3]

Sometimes strange remains are found in the oddest of places.

On 11 June 2018, Duncan Mackay, a crofter in Rogart, Sutherland, got a surprise when a digger extracting peat on his land unearthed a number of dark, tanned bones surrounded by a mass of curly golden red hair at a depth of around 1.5 metres. He phoned the chairman of the Clyne Historical Society, Nick Lindsay, who came to view the remains the next day. While Duncan had initially thought, given its colouring, that it must have been a fox, cursory examination of its skull ruled out this explanation. Nick collected all the bits, put them into plastic bags and sent the remains to Inverness Museum.

A few days later, what was by then being called the 'Rogart Bog Beast' on account of its sharp teeth was assessed by Jeanette Pearson, the museum's conservation officer. At first glance, she ruled out any possibility of fox and suggested that, as its skull was roughly half-way in size between that and a wolf, it must be a big dog or a young wolf. Young wolves, which are dark at birth, become lighter as they age. At an age of around six months, they can be relatively rufous on their legs, flanks and face, with a soft creamy grey background colouration elsewhere. But their fur is neither long, curly nor golden overall.

Subjected to further examination, it became clear that the Rogart find was like nothing else ever known to have lived wild in Britain. Was it just a dog? I spoke again to Nick, who stated that had it been a dog, he could see no reason why anyone would ever have bothered to have buried it that deep. As peat formation occurs at a rate of around 1 millimetre per year, if the creature had died on the surface or sunk below into perhaps a shallow mire, it could have lain there since around 500 CE. The possibility that it had burrowed was impossible to prove as the digger had removed the surrounding ground.

While peat, depending on its character and pH, can change the colour of fur, Dr Bryony Coles, the great historian of *Beavers in Britain's Past* and who has undertaken extensive excavations in wetland environments, is of the opinion that when it goes in looking one way it generally comes out looking the same. So, the Rogart Bog Beast, of whatever sort it was, was golden brown when it lived. While its remains await DNA analysis, the skull has been carbon dated and is approximately 150 years old.

It's most likely a dog.

Although genetic evidence for this is currently unclear, wolves may have dissolved into dogs in Britain.

Most Likely a Dog

Sir Henry Hamilton Johnston was a British explorer and geographer. In 1882, he went to Angola with the 7th Earl of Mayo, Dermot Robert Wyndham Bourke. They got on well and, as Bourke owned extensive lands in County Mayo, he introduced Johnston to Achill Island on Ireland's west coast. The island impressed Johnston so greatly that he was wrote to *The London Times* in December of 1903 that 'all that is wanted to effect the end of making Achill a paradise is to constitute the island a national park.' In his letter, he noted that 'some of the domestic dogs on Achill … exactly resemble the wolf in colour, in brush, in the shape of the ears, and in the arrangement of the masses of hair along the line of the back.' It is believed that these little wolf dogs survived until well into the 1920s, though no trace of them can now can be found.[4]

In the late 1800s, amateur archaeologist John Beecham discovered the entrance to a small cave on a scrub-covered part of Lakeland hillside at Helsfell Nab near Kendal. Over the summers that followed, he spent his evenings digging. Having worked 'with dogged pluck for five continuous summers', he was rewarded by the discovery of the remains of bear, fox, wildcat, polecat, otter and aurochs.[5] When he could literally dig no more, Beecham resorted to the ultimate Victorian excavation tool, dynamite, and when the dust settled he found the skeleton of what he believed was a wolf behind a large rock at the back of the cave. His description of what he found was pure Victorian pathos: 'The poor brute,

149

feeling no doubt, that its end was approaching, had retired behind a large rock and there had died in self-imposed solitude. Its fellows respected its remains, which remained undisturbed until the explorer discovered it in situ, the bones lying close together.'

It's on display now in Kendal Museum, where I went to see it in the summer of 2021 on a day when the museum, despite stating on its website it was open, was patently not. After I'd knocked in an undeterred fashion on its front doors, a couple of charming chaps from a local college appeared to ask what I required. After I explained, they said that, while they would gladly show me the wolf, they had no idea where the light switches were for the gallery. The only illumination possible therefore would be from our phone torches. As a spookily atmospheric opportunity it was memorable. Under the gaze of large African ungulates peering down from the walls with fixed frowns, we checked countless cabinets of dried puffer fish, South American spears and old leather shoes. Suddenly, next to a rigid beaked eagle with unfeasibly outflung wings, there it was.

A sort of whippet-sized wolf.

While radiocarbon dating suggests that the Helsfell wolf died between 1139 and 1197, its DNA has not yet been either collected or analysed. If it's a wolf, it's a slight one.

There can be little doubt that the Welsh once knew their wolves well. They named their finest horses after wolves when they moved with their fluid grace and nurtured apocryphal sayings such as, 'worse a wounded wolf than two healthy ones'.[6]

At the foot of the castle at Laugharne on the south coast of Wales is Island House, which dates back in time to the sixteenth century. Around 1880, when it was sold, the former owner removed some artefacts that had been hanging under the open chimney in the great hall for 'all time' and gave them to the vicar of Laugharne for safe keeping. These included two wild boar skulls, the horns of a roe buck and four wolf paws. These were placed in an old chest and forgotten about until 1926 when the island was inherited by Anthony Congreve. He was interested in hunting and, on learning of the relics, asked for them back. By that time, some had gone missing and only one boar's skull and three of the wolf paws remained. In 1928, he sent the skull and paws to the British Museum for examination by a Mr Reginald Innes Pocock, who commented:

They are shrunken and entirely bare, not a trace of hair remaining. As far as size is concerned, they may have belonged to a wolf; but, since there is no structural character by which a wolf's paw can be distinguished from the paw of a large dog, their precise identification is a matter of conjecture. Since however the spoils of the Roe Deer and Wild Boar strongly support the tradition that they are hunting trophies dating from the time when these animals still roamed the woods of south west Wales, I see no reason for dissenting from the tradition that these pads are those of a contemporary Wolf. If this be so, the set of trophies is, as far as is known, unique.[7]

In a booklet about the history of Llandeilo by Eirwen Jones, published in 1984, is a picture of the wolf paws from Island House with this caption: 'Wolf pads found at Dinefwr'. The author offers no explanation in her book about this alternative assertion and it is not clear if there was any evidence that the paws had previously come from Dinefwr Castle before being moved to Island House. Roe deer are known to have survived until the sixteenth century in the area, and the two boar skulls are also of interest given that oral tradition suggests that the last of their race was despatched at Derwen, some two and a half miles to the south of Dinefwr, at around the same time. Anthony questioned local people, who informed him that the wolf was believed to have been killed at Panteague in the parish of Marros, about seven miles to the west of Laugharne, although when and by whom was not clear.[8]

The three paws that remain, two hind and one fore, are now pinned to a wooden shield in Museum Wales. From correspondence in its archives, it appears that Anthony had them mounted, as they were loose when he received them.

They are big.

The hind paw on the left side of the shield measures 230 millimetres from the cut edge of the skin to the claw tip and 48 millimetres across the metatarsals. The paw on the right is 210 millimetres and 48 millimetres respectively. The forepaw in the centre measures 260 millimetres in length and 50 millimetres across the metatarsals. Donated to the museum in 1977, they remained in storage until 2002, when Professor Luca Fumagalli from the Institut d'Ecologie at the Université de Lausanne, who was conducting research into the genetic variation of European wolves, sought samples from historic specimens. He was able to extract mitochondrial DNA from a section of skin taken from the Island House paws, and wrote to the curator at the time to clarify that the DNA sample he had obtained matched very closely with other wolf material already in his collection that originated from the geographic area of western France.

As we have seen, fox hunters from Britain regularly attended wolf hunts in Brittany and that they could have been obtained there is undeniably the case. It is also possible that they could have come from a pet. As the issue of whether the paws dated from a period around perhaps the mid 1600–1700s, when a few wolves might still have conceivably roamed Carmarthenshire, was unclear, a bone sample was submitted by the museum to Oxford University in 2008. The result that came back was a probability of 27.6 per cent that the actual date of death was between 1682 and 1736, and a 67.8 per cent chance of it being between 1805 to 1934.

If you go for the lower percentage, then evidence from elsewhere suggests that they could have come from a British wolf. If you plump for the latter, then it's much more likely they have come from elsewhere.

The paws now are completely hairless like long, dry evening gloves. There is nothing in their size or shape to suggest that they did not all come from the same animal. All the claws are present and these have one last story to tell. While they all have signs of wear on their tips, one outer claw on a forepaw is broken.

The fractured edge of the stub that remains shows that it incurred wear before death.

Did it tear when the last hunt began, before smoothing as its owner raced and scrabbled in a chest-bursting, fear-filled chase before the hounds in hot pursuit tore it down and it was spliced on the lance of a lord?

———

Ben had kept his bones for five years in his bedroom and the obvious next step was to arrange their sampling for radiocarbon dating and DNA. He agreed to this and a tibia was sent to Danielle, who forwarded it to a laboratory in Florida. After months of waiting, they replied to inform us there was a problem. They attached an image of something that looked like a tiny drop of spun candy floss, which was the collagen they had obtained and expanded from the bone. The small black areas in their photograph were the tips of plant roots, so tiny as to be invisible in the pretreatment process. They advised us that if these were of a different age to the collagen, they would skew the result to more recent times. There was no option other than to pick them out of the sample in the laboratory and continue on the basis that they avoided any visible roots.

We waited again and on 19 October 2020 the results returned.

They were from 1980.

At a time when no wolf could have been there.

Terry and Ben were deeply disappointed and we reconsidered our position. Was it possible that the bone collected from the location next to the skull was from another era? Ben thought not, but another friend who was an archaeologist of the medieval period informed him that he had experience of material from one of his studies returning erroneous carbon dates from the 1950s or 60s. Could this have been the case with the tibia?

The only solution was to test the skull but, before doing so, Terry asked Danielle if she remained of the opinion that it was a wolf.

On 4 November, she replied:

There is nothing about the skull I have seen that leads me to conclude it is anything other than wolf. Morphologically and metrically, the postcranial remains are wolf. There is no reason to suspect the bones are not from the same animal as the skull even though I know there are two individuals and I haven't been shown both of them.[9]

Terry sent Danielle the second skull from the cave entrance. She examined it and produced another surprise:

> The second specimen is a dog, no question about it. Terry says the skeleton was near the cave entrance, whereas the first one was at the end of a long right-angle burrow into the cave and could not be reached without being dug into … It's not an issue anyway to have wolf and dog in the same system since they would have overlapped/coexisted for thousands of years.[10]

After further discussion, we decided to send another sample from what Danielle believed to be the wolf to Greger Larson's laboratory at Oxford for genetic confirmation. While the date from Florida seemed just too incredibly recent to be true, we wanted to solve this mystery very much and no other avenue of exploration remained.

The first skull that we dated from the burrow was a wolf, while the other undated at the entrance was a dog. Did they live alongside each other? Were they friends? Or mates? Perhaps the wolf took solace, when all its kin had gone, in the company of a stranger of a different form that was near enough its understanding for comfort?

— CHAPTER EIGHT —

Severing Their Umbilicals

Not fewer the skulls, cold and empty,

.   .   .   .   .   .

I see no stake without head;

.   .   .   .   .   .

On sharp pointed hazel staves.

Giolla Críost Táilliúr, 'O Trinity, Bless Your People'

In 2003, I bought a small farm in France. It was a joyous mistake. Near the monastery of Mont Saint-Michel, high on its rocky outcrop I retained an elysian vision of retiring to run years of sun-filled wildlife tours, cultivate vegetables and rent gîtes. This Rosé-d'Anjou-tinted dream was not to be. It became a needless burden beset with long boat journeys through the night to mow tidal waves of brambles, which in the end engulfed the property to the dismay of the neighbours. In the end, it burnt time I did not have. No amount of cheap wine could compensate for the stress involved and, in 2008, I sold it, but not before I realised that a nearby lane was named 'Le Loup Pendu' or 'hanging wolf road'.

Viewed as criminals by society at large, wolves were often hanged. In the time of Edward the Confessor (1003–1066) *'wulfhēafod-trēows'*, or wolf-head trees, grew in the forests of Britain. Ending their lives in this way was a popular pursuit, and Paul Sébillot recalled seeing a wolf hanging from the branch of an oak tree in Brittany around 1860. In 1884, a large number of people came to see the body of a wolf hanging in the covered market at Jugon in the Côtes-d'Armor, and, in nearby Rohan, if a wolf was caught when its princess was out hunting, the town's cobblers traditionally were required to drag it through the town before stringing it up to the 'wolf oak. Dead or alive'.[1]

While 'Dog Hanging' wood in Bringewood Chase near Ludlow still commemorates the fate of canine transgressors against the king's deer, no wolf-hanging trees in Britain are known to survive. Still in existence, dole, dule or dool trees, whose name is drawn from the old Latin term, *dolor*, for pain or grief, are where human executions took place, like an ancient sycamore of nearly a metre in diameter in the grounds of Leith Hall in Aberdeenshire. On a recent visit to the Graythwaite Estate in the southern Lake District, the forester who kindly showed me round offered to give me a seedling from the last of the dole oaks in the north, which had blown down two years before. I have yet to collect it but have every intention of doing so soon.

When it gets to an appropriate height, I intend to use it for hanging my bird feeders.

The effort made by our forebears to extinguish any form of wild creature that has opposed our interests in Britain is near complete. Subjugation was never enough and the larger creatures that survived into modern times in plenty – hares, foxes, badgers and deer – have done so only because they were carefully cultivated for the sole purpose of hunting. If we did not kill them outright, we retained them to kill at a time when it suited us better, wearing comedy costumes and parping ridiculous trumpets.

Any other creature not fitted for pursuit of this sort was worthless and we strained every sinew for centuries to sever their umbilicals and extinguish them forever from the earth.

The wolf, because of its demonisation, became for a time our firmest of foes and as such was treated by us in a manner that we now know to be much lower than bestial.

In the second half of the twelfth century, a draughts-man of carved walrus ivory from the medieval school of St Albans illustrated two men holding an upside-down creature with the features of a wolf. Although it may have been killed already, it's just as likely they were readying it for their pleasure.[2]

In 1814, the naturalist John James Audubon watched a farmer deep in the Ohio River Valley hamstring and torture three wolves he had caught in a pit trap. He was 'paying them off in full' for taking 'nearly all of his sheep and one of his colts'. While he had no evidence that his captives were culpable, all their kind were wrong. What surprised Audubon most was not the nature of their end but that, once disabled, they 'lay flat on the earth … ears laid close over their head, their eyes indicating fear'. One large dark male was so intimidated that, when hauled to the surface, it remained 'motionless with fright as if dead, its disabled legs swinging to and fro, its jaws wide open, and the gurgle in its throat'.[3]

Captured wolves with their hamstrings cut could be set against dogs. Even when disabled, they made dangerous opponents. Louis XIII of France acquired an old wolf and set his best dogs on it, three at a time. The wolf managed to kill twelve of them without sustaining any serious injury. In 1597, Thomas Fenton, who was the keeper of the royal menagerie at Holyrood, and as such was already in charge of its lion, a tiger, and other exotic animals, was given custody of 'two Wolfis' for which he was only paid 40 days' keep.[4] While they might have been obtained for display, it is entirely possible that their numbered days were also predetermined for a spectacle of this sort. The fighting styles of wolves and dogs differ significantly, and while dogs typically limit themselves to attacking the head, neck and shoulders of their opponents, wolves make greater use of body blocks and seek control of their opponents' extremities. Theodore Roosevelt once wrote that he considered a large mastiff to be a match for a wolf only if the latter was 'a young or undersized Texas wolf. Even if the dog was the heavier of the two, its teeth and claws would be very much smaller and weaker and its hide less tough'.[5]

One English witness to a wolf captured in a Breton pit in the mid 1800s stated that, 'The ... beast was raised from its confinement by means of a clasp fastened to a pole, and when helpless in mid-air it was a simple matter to tie its paws together with running nooses. Once out, its head was pinned to the ground with a fork while its lips were sewn-up with shoemakers' thread to muzzle its terrible jaws.' The author then added as a thoughtful after-note, 'That such methods may seem cruel but the wolf itself is a cruel beast and could scarcely expect to be treated better.'[6]

Another description from Scandinavia of how the wolf pits were used states that any captured wolves were first bludgeoned to subdue them before being bound and taken alive to the villages, where they were slowly and 'voluptuously' tortured to death. Although I am not sure that the word voluptuous is correct in translation, I have come to consider its use well founded as it describes quite exactly what occurred. Revelry and pleasure in the end of an enemy. Clubbed but not enough to kill. A drawn-out death of many cuts, strokes, tugs, tears and wrenches. Every cry, shriek, whimper and contortion enjoyed. After considering this dark trajectory, I wondered if wolves were physiologically capable of producing tears. Romain Pizzi, a respected international veterinary colleague who has worked in the harrowing environments of poor zoos, bear bile farms and wildlife rescue centres worldwide, said initially that he was not sure that the answer would be known. After some thought, he called back days later and suggested I speak to a contact of his who dealt with dogs rescued from restaurants in China.

I have not yet summoned the courage to ask – and don't think I ever will.

───────

Wolves could be snared in a multiplicity of manners. In early times, a deadfall of rock, ice slabs, tree trunks or snow could be utilised to smash any puller of bait balanced on a trigger of cord. Another method in Germany was to construct a dense hedge with a few select holes into which snares made from wire loops were set before the wolves were chased through them. They quickly learnt to avoid these hedges and as a solution this did not work well, but toothed steel-spring traps, bow traps, which crushed and smashed heads and ribcages, spear traps or sprung bolts could all be alternatively employed.

Wolf pits were an effective form of capture. They still litter Britain and, like those on Roy Jones' Wolfpits Farm in Wales, were often created in clusters to ensure a good haul. Although some remain obvious as marked locations, others exist in dioramas of disguise. Feveras, Woolpit, Ulf pits, Wolf folds, Wlfpit, Wlvelei and maybe Luppitt (loup pit) in the south-west were all locations of pit traps.[7] Masked by Old Cornish, which called wolves *bleit* or *bleydh*, Lower Blakefield, Higher Carblake and Lower Carblake recall their presence throughout what was once the wilderness of Bodmin Moor. On Exmoor, where even the furthest memory of the wolf is near extinct, it is only in the roll call of the old royal forest farm lands that the name of Woolpitland endures.

The Reverend James McLauchlan of Dalrossie Parish observed in 1835 that the method of capturing wolves in pit traps was as follows:

Half way across the mouth of the pit a broad plank was projected, about half of which lay upon the ground, and had upon the end farthest from the pit a weight sufficient to balance a wolf ... on the other end was placed a bait, and the remainder of the pit was covered over with brush-wood, so as to deceive the animal. The wolf advanced to the bait along the plank which, when he overbalanced the weight on the other end ... and he was precipitated into the pit.[8]

Elsewhere in the 1800s, a round pole was knocked down into the centre, to the top of which a hen, or in Brittany a quacking duck, was usually tied. Narrow rods were then placed over the pit and on top of these a thin layer of spruce twigs or moss. Once the covering was complete wolf dung, if it could be found, might be thrown over the surface, while other baits, such as dog scents or the contents of fox bladders, could additionally be used.[9] In this way, wolves and other predators would be encouraged to investigate and then plunge through. While the bottom of the pit might be filled with iron spikes or sharp wooden palings to ensure that any falling creature might be impaled, there is contradictory evidence regarding this practice. No killing in or next to a pit was allowed in Scandinavia as it was believed the scent of fear it engendered was so pervasive that other wolves would avoid the location. The best time for setting newly made traps was reckoned to be in March or September in the belief that during those months wolves could not sense newly excavated soil.

More crudely, on the ridge lines in the remote upland landscapes of Britain, which wolves were likely to follow, traps made from rock slabs on their tops, sides and floors were installed. These structures operated on the basis that a swinging wooden door at one end that only pivoted inwards would enable predators to crawl in to access the bait within. Filled with carrion, they lured in all sorts of the unwary, which, once secured, if they had no worth might be left to accumulate like slugs in a beer glass.

Needles or hooks called 'housepieds' inserted into chunks of meat and scattered at random killed everything. Once lodged in their consumers' mouths, throats or bellies, the end would be slow, lingering and inevitable. Early settlers in the Americas were supplied by English merchants with 'wolf bullets and adders' tongues' – another form of needled spring trap – and large wolf hooks, which were baited and suspended from a rope on a bough set just high enough for a wolf to jump up and grab.[10] One tug and sprung steel callipers would smash into their muzzles to grip like a vice.

Another vile contraption was the 'wolfsangel', which comprised two metal parts and a connecting chain. Its top part, which resembled a crescent moon with a ring inside, was fastened between the branches of a tree while the bottom part, onto which meat scraps were attached, was a hook. Once swallowed, even if the wolf struggled free, the chain would catch itself on obstruction after obstruction, ripping and rupturing with every wrench. No escape was possible.

Poison was also good. While many could be killed quite readily with carcasses scattered at random, a potion well made could be smeared on arrow or spear tips to ensure that a graze or slight wound would still prove fatal. Recipes for the preparation of a brew could be made from the roots of the white hellebore for game of the 'most choleric temper', like the boar, the wolf and the wildcat, or from the attractive and rare cream-petaled plant with brown spots, which flourishes sometimes on the disturbed soils of construction sites, called 'hen-bane'.[11]

Although the strychnine tree, which has been known since ancient times as 'nux vomica', is native to India and Southeast Asia, its seeds were widely traded. One French recipe that employed these involved a pound of ground seeds in combination with twelve bulbs of wolf-bane. Wolfbane is an aconite that can be found growing wild in the herb-rich pastures of high alpine meadows. On account of its pale-yellow flower, it was once known as the white-bane to distinguish it from the deep indigo of the domesticated form, which is otherwise known as monkshood. Presumably the dark, elaborate cowl of the latter gave that form its name, although the monks themselves were dedicated killers. The Hungarian name for the wild form is *Frarkasolo* or 'wolf killer', while the cultivated plant is called the 'wolf's aversion' in Gaelic and in Welsh 'wolf choke'. The pound of nux vomica together with the aconite bulbs was then ground in a mortar before a final infusion of nux vomica mixed with horsehair was added. When this process was complete, a sore-ridden village dog was obtained and stabbed to death. The holes in its flesh were stuffed with the potion before its carcase was inserted to rot in a dung heap for several days. Once suitably stinky, all that remained, bloated and blueing, was removed to a place where the wolves might occur.[12]

In the end came the bounties. If you want people to kill other creatures or other people, history shows you must always pay well. Although under Henry II (1154–1189) wolf heads were worth just a few pence, by the time of King John (1199–1216) they had risen in value to 5 shillings.[13] In 1458, the bailie of the earldom of March paid Gilbert Home 5 shillings for killing ten wolves in Cockburnspath in Berwickshire, and in 1498 the lords of council at Inverness enacted that, if anyone brought a wolf's head to the sheriff, the sheriff or bailie was to see that he received 1 pence from every five houses in the parish. By 1491, a man at 'Lythgow was paid [5 shillings] ... for two heads'. As wolves grew scarcer, the prices went up and payment of 'six poundis threttein shillings four pennies gieven ... to Thomas Gordoune for the killing of ane wolff' was awarded in Sutherland in 1621.[14]

In 1652, in Ireland an order was issued that Richard Toole with Morris McWilliam, his servant, their two fowling pieces and half a pound of powderman's bullet portion should be permitted to pass quietly from Dublin into the counties of Kildare, Wicklow and Dublin for the killing of wolves. This instruction was to continue for the space of two months from the date of the order.[15] At the first political meeting called by the Parliament of England to assure an Act of Settlement for Ireland in June 1657, Major Morgan for Wicklow stated that, 'We have three beasts to destroy, that lay burdens upon us. The first is the wolf, on which we lay five pounds a head if it's a dog, and ten pounds if a bitch. The second beast is a priest, on whose head we lay ten pounds; if he be eminent, more. The third beast is a Tory, on whose head if he be a public Tory, we lay twenty pounds; and forty shillings on a private Tory.'[16]

For even the lesser wolves there was no element of jest, just a literal list of tariffs. 'For every cubb which preyeth for himself, forty shillings; for every sucking cubb, ten shillings. And no Wolfe, after the last of September until the 10th January be accounted a young Wolfe.'[17]

While the professional wolf hunters lured out to Ireland with incentives of this sort and gifts of land had to kill wolves or face fines, bounties that pay readily can easily turn inwards. Once something is valuable, it's worth producing. In lost church records from Flixton and Saxton in Yorkshire, it was commonly said in the 1880s that the parish registers contained notes of payments of bounties to 'wolfhunters ... the sum of 5 shillings a scalp'.[18] This is a very late date and the tendency of parish clerks to take records with them when retirement came or to destroy papers they considered of little worth means that no exact detail of what was paid for can now be found.

Were the claims errors or for dogs that were just wolfish enough in appearance to be delivered up for loot?

Corruption is an eternal human quality.

While the unflinching application of the methods described above erased the wolf, the same savage strategy of oppression turned our whole island into one great killing ground. Every small creature that displeased had to be scourged. Recently, on a trip to buy wildflower seeds from a local farm to restore a diversity of plant life in my meadows, I came across a goodly bunch of gin traps hanging from their chains of wrought iron from a beam in an old cart shed turned into a classroom. When I commented on their presence, the owner remarked that these were only a fraction of what he had found in the buildings when he bought the old farm of some 60 acres.

Set among other farmyard paraphernalia of a bygone time, they were rusted and worn but the small brass clasps that set their treadles to spring would have surely worked if you oiled them. Some were small, for weasels or hedgehogs, with larger versions of a similarly square-jawed design for polecats or hares. Beyond this, in circular pattern with a football's circumference and jagged, sharp teeth in their rim, were deep biters of badger bone.

In 1904, a wolf escaped from a small zoo near Hexham and began to kill sheep. Hunts with rifles and shotguns were organised and a committee chaired by the local MP, Major Wentworth Henry Canning Beaumont, assured crowds of torch-bearing farmers in market squares that it would soon be brought to justice. On 29 December, the body of a large male wolf, which had been cut in two by a train, was found on a railway line and brought back to Cumwhinton station. Pictures of its bloody carcase still adorn the walls of many Northumbrian pubs where beery conversations regarding its grisly depredations continue to this day. As nothing much else ever happens in the north, a ceremony to commemorate its passing began in Allendale in 2014 wherein mouth-moving wooden wolf effigies, which have featured 'grandma wolf' and 'Dalek wolf', are now annually burnt to the delight of large assembled crowds.

Ambrose Barnes, or 'Roast Wolf' to his friends, burnt wolves as well. Real ones… alive. His nickname suppos-edly was awarded as a result of the many wolves he had killed and occasionally eaten during the reign of Henry

VII (1509–1547). That wolves are edible is a truth. Very little lives that is not, but in an illuminating interview on 20 January 2020, Glenn Villeneuve, of reality TV series *Life Below Zero*, described how he had occasionally consumed them. Despite prompting from a grinning TV host, he emphasised this practice as being born of necessity rather than culinary enjoyment and gave no impression whatsoever of enjoying his repast.

My Uncle Tommy had been a giant in his day. When I was small, he had been attacked by a Lincoln red bull and, with the help of his wife, Annie, fought it off with a yard brush. It was sent to make sausages. The last time I saw him before he died, in the tiny village of West Linton near Edinburgh, we discussed sheep and cattle comfortably. I was farming at the time and that talk passed with ease. He asked about my beavers in a polite sort of fashion before remembering the wolf cubs.

Quite suddenly, he asked me to promise that whatever course I pursued in life I should never support any action that assisted the return of the wolf as they were such very cruel creatures.

He was a nice, kind man. If Tommy had ever seen a wolf, it would have been a broken old beast in a zoo.

CHAPTER NINE

Void of Noisome Beasts

Here's thunder on the Blorenge,
Hark! echoing far it sounds
O'er fair Llanover's sloping sides,
And Goytrey's woody bounds;

. . . . . .

A hollow, wailing, long-drawn cry,
The Gwentians know the tone:
The last old wolf, his race all slain,
Howls on the hills alone.

Charles Hanbury Williams, 'The Last Wolf in Gwentland'

Greger Larson's laboratory responded with their result regarding the wolf skull from Yorkshire in the summer of 2022. While an initial intriguing email from his assistant suggested that it was either a wolf on a dog's genetic spectrum or a dog on a wolf's, later clarification confirmed it to be a probable dog with a radiocarbon date of around the 1920s. This finding prompts the question of how two large dogs could have lived alone in a cave, gathered carrion to feed, perhaps attacked domestic livestock and dug wolfish burrows in its substrates so very recently.

We know that big dogs have turned feral in the past. In 1784, one that was abandoned by a fishing vessel near Boulmer on the coast of Northumberland began to kill sheep. So effective did it become that it operated within an area of around twenty miles surrounding Heugh Hill, near Howick, where it lived in a rock den. Once it caught a victim, it would rip a hole in the sheep's flanks to enable the consumption of the tallow round its kidneys. Incredibly, as in my experience dog bites on sheep swiftly become putrid, several of the ewes survived this ordeal and, well cared for by their shepherds, subsequently gave birth to lambs. Farmers who sought to destroy the dog used hounds but, when these caught up with it and it rolled on its back in suppliance, they lost interest before their pack of panting owners got within shooting range. The dog was eventually killed in March 1785 in an ambush at its cave.

Dogs of this sort that commonly killed sheep were widely called wolves. Although Ben's skeletons in their Yorkshire cave were large, there was no surviving fur or any other clue regarding their appearance. As no physical evidence identifies their cause of death, perhaps poison was possible?

It's unlikely we will ever know.

Void of Noisome Beasts

The English often insisted, as William Harrison the Canon at Windsor did in 1577, that:

> It is none of the least blessings wherewith God hath endued this island that it is void of noisome beasts, as lions, bears, tigers, pardes, wolves … This is chiefly spoken of the south and south-west parts of the island. For, whereas we that dwell on this side of the Tweed may safely boast of our security in this behalf, yet cannot the Scots do the like in every point wherein their kingdom, sith they have grievous wolves.[1]

Statements of this sort were, however, bumptious. Not so many years before, in 1538, a French chronicler, having been assured in a similar vein that no wolves remained in England, recorded the sighting of one to the south of Berwick.[2]

An absence of wolves was an issue of nationalistic English pride. They were the civilised nation on the island, and the others to the north and west, let alone the savages that existed on the other side of the Hibernian Sea, were not. When new settlers from England or Scotland arrived in Ireland, they referred to their new home as 'Wolf-Land' and thought it to be primitive and low. This understanding extended to its people, who were considered so primal that one Captain Ireton reported to his commander in 1647, after his men had slaughtered the garrison of Cashel, that many of the battlefield dead he had personally examined had tails.[3] All was quite simply bestial when you left settled England.

For as long as wolves endured somewhere on the main isle of Britain and, in the absence of any physical borders between England, Scotland or Wales, they would have returned on occasion anywhere else. The scattering of 'last wolf' stories that occur in Wales and England in landscapes long believed to be too well settled and thus inhospitable to them could simply be explained by the movement of cubs from the far north exploring in search of mates and suitable living space. Although past authors have inclined towards their extinction by the late 1600s, there is evidence that suggests they lingered longer.

The last wolves took a long time to die.

Before we begin their stories, it's worth understanding that many, if they were not fables initially, have with time transformed into something very similar. While this may be a result of the oral heritage of historic recollection that was normal in Gaelic and Welsh, even written English history is much less than a thousand years old and, as a result, the exact dates of many great events remain unclear. Given this lack of detail, any absence of clinical exactitude regarding the deaths of a few wolves is hardly surprising.

So, let's start with the tall type of stories, with no timeline beyond long, long ago.

One day, two doughty Scots women were chatting on a cairn after the first had gone to her neighbours to borrow a griddle when they heard beneath them a wolf...

The bowels of the carn, however, were so straight and intricate, that as he dragged his lean body beneath the holes at the feet of the two gossips, they became aware of the scrapping of his claws, and the trailing of his gaunt limbs through the rocks and withered leaves, and as their eyes turned to the ground, they met the round glazy balls of the wolf glaring up out of a chasm amidst the blaeberries.[4]

One of them whacked it on the head with the griddle – without denting its pancake-producing potential – and the wolf promptly died.

In the South West of England, the last wolf in Cornwall lived in the forests of Ludgvan, near Penzance. As the only member of his race, he was a gigantic specimen, capable of playing terrible havoc with sheep flocks. Tradition tells that at last, one day, he carried off a child. This could not be endured, so the peasantry turned out and killed the wolf close to a farm called Resparveth, which exists still to this day.

The Bleeding Wolf Inn in Cheshire on Scholar Green is a Grade II-listed public house built in 1936 for Robinsons Brewery, replacing an earlier hostelry on the site. Local legend has it that a forester called Lawton saved King John in the early 1300s from being killed by a wolf at the spot where the inn now stands. As a reward, the king gave the forester all the land he could walk over in a week starting at the wolf's body. Lawton took him at his word and covered enough land at a trot in the time allowed to establish a great estate of his own.

When Edward Postlethwaite wrote a 'last wolf' poem in 1496, his drama regarding its death played out with great elan on the cliffs of Humphrey Head in Cumbria. In a ballad reworked by other writers since to give it historic punch, it was given again by Mrs Jerome Mercier in her 1906 book *The Last Wolf,* where it ran to seventy-five verses.

The story begins when Sir Edgar Harrington of Wraysholme Tower gets up one morning and decides on a whim to hunt and kill the last wolf in Cartmel Forest that very day. He promises that the man who succeeds in this task will receive the hand in marriage of his niece, before bribing any likely participants who may be wavering with further offers of land and money. The noble huntsmen, thus incentivised, assemble swiftly and scare the wolf from its lair on Humphrey Head before chasing it as far as Windermere, where it swims the water body. Luckily, when the worn-out wolf returns to Humphrey Head after its long day out, a mysterious knight who turns out to be Sir John Harrington, Sir Edgar's long-lost son, is just in time to behold 'a young and lovely female, crouching in a cleft of the rock' while the 'enormous wolf was endeavouring to reach her'. Sir John transfixes the wolf on the tip of his hunting spear and is gratified to discover that the lady – Edgar's niece and his own very close relative – has fallen in love with him, at which point they marry and live happily ever after.

The wolf symbol that Sir John assumed for his coat of arms in the form of a golden weathervane flaunts the skyline atop the church at Cartmel to this day. His effigy and that of his bride wrought in marble lie inside with the wolf – a gremlin-pug sort of effigy – spread across their feet.

While every parish in every county and every shire once had its own 'last wolf' story, the historic records of these no longer extend beyond a mere few words. In 1328, a wolf was killed in the Royal Forest of Clarendon near Salisbury and when Ralph Higden, writing in 1340 from Chester, commented that there were now 'few wolves left in England' he was quite simply mistaken. There are records of last wolves at Marske in the North Riding of Yorkshire until 1369, and John of Gaunt, the Duke of Lancaster (1340–1399), whose vast chase stretched over 250 miles from Wensleydale to Stanemoor, was supposed to have killed his own last wolf in a finale during a hunt in the parish of Rothwell near Leeds.[5]

William Somerville and Edward Topsham's 1817 book *The Chase* states that the last place in England to put a price on the head of the wolf was on the Yorkshire Wolds. And the village of Perry Oaks, which is now underneath Terminal 5 of Heathrow Airport, was 'once an oak forest that spread its green boughs from Staines to Brentford ... there is an old tradition that the last wolf in England, killed centuries ago, was hunted down at

Perry Oaks'.[6] The village of Barthomley in Cheshire had a wolf killed in the fifteenth century, with a local brook, the Wulvarn, being named in its memory. In Derbyshire it was Wormhill, in Cumbria it was Great Salkeld and in Sussex Wolfscrag.

In 1592, Frederick the Duke of Wurttemberg recorded a wolf in the Tower of London, which had been previously described in 1589 by Paul Hertzog as being 'extremely old'. Frederick noted it was, 'a lean and hungry wolf, which is the only one in England; on this account it is being kept by the queen – and indeed there are no others in the whole Kingdom, if we except Scotland, where there are a great number.'[7] No other wolf of the many that subsequently occupied the old menagerie at the tower was ever referred to again as having any English connection.

Factual records concerning wolves in the Highlands are not uncommon in the sixteenth century, although they may have become harder to find towards the end of it. Some records were of a prosaic sort. No last fights on mountain passes. No lances or swooning maidens. In 1593 a clerk responsible to the Breadalbane family recorded that a flock of sheep, stolen in Lowland Perthshire, were brought to Killin, where 'the wolf slew six of the same',[8] and a year later, in 1594, a two-year-old quey (young cow) and four horses were killed by four wolves in another two incidents nearby.[9]

In the early 1620s, the tenants of the Breadalbane lairds were still honing four wolf spears annually as part of their feus. Though Sir Robert Gordon of Straloch observed in 1630 that 'Noxious animals and such as prey upon flocks are absent, except foxes, and these are rare, for wolves are believed to be now all but extinct, or if any exist, they are far away from the more cultivated localities and human civilisation', Sir Robert Gordon of Gordonstoun confirmed in stark contradiction at the same time that there was still good hunting in Sutherland as its forests and chases contained an abundance of deer, wolves, and other fauna.[10]

In a debate before the English House of Lords in 1641, the solicitor general, Oliver St John, remarked, 'We give law to hares and deer, because they are beasts of chase; but we give no law to wolves and foxes, because they are beasts of prey, but knock them on the head wherever we find them,.'[11] While we can accept with ease that hares, deer and foxes were still present in his time, he made no distinction nor applied any past tense to the wolf. As late as 1680, the owners of Llidiardau Castle in Wales, who wished to keep a wolf in their moat, recorded that their agent 'Payd the man that came from llewnie with the wolfe 2s.-0'. Subsequent correspondence that a 'Mr Bell Jones has informed me that you have an inclination to have a wolfe and I have a very fine one that is pretty tame' followed thereafter. Other records clarify that further wolves were maintained at the castle until at least 1723.[12] Within this timescale, it is possible that at least some of their captives could have been obtained in Britain.

Hunt for the Shadow Wolf

Furthermore, after centuries of snootily reaffirmed assurance that wolves were no longer to be found in the civilised south, another writer stated in 1691:

And yet I am credibly informed, that in some Places, as Warwickshire among the rest, some wolves from time to time have been discovered. But, as it happens but seldom, so upon the least notice the Country rises amain, as it were against a common Enemy; there being such a hue and cry after the Wolf, that it is hard for him to escape the Posse Comitatus.[13]

For those of you like myself who no longer speak Latin, the *posse comitatus* is the power of a county. It's the force of its entire body of inhabitants, who may be summoned by a sheriff to assist in preserving the public peace or in the execution of legal precept. It applied in old England under the common law to every male inhabitant above fifteen years of age and any wolf wandering through Warwickshire would have been lucky to have escaped with its life. But if odd wolves were still to be encountered in that settled county at the dawn of the eighteenth century, where were they coming from and what were they doing in wilder landscapes? Surviving, it would appear in the far north, if Polson's story from Brora from around 1700 and MacQueen's fight on the Findhorn are to be believed.

When Lord Morton was president of the Royal Society in 1756, he told the French naturalist Comte de Buffon that there were still wolves in the Highlands,[14] and in

1760 a wolf was supposedly killed in a gully above a farm called the Shenval of Gairn in the Grampian Mountains by a farmer's son with a bow and arrow.[15] Shortly after, in 1763, a wolf was supposedly killed in Caithness near Berriedale in a hunt organised by the Hendersons of Stemster. The jawbone of this beast that was retained by the family as a trophy was, however, identified in 1951 by Professor Ian Ware as positively belonging to a Bengal tiger, and this account must therefore be consumed with a very large pinch of salt.[16]

Near Nantmel Rhayader in mid Wales, the Afonwy foxhounds supposedly brought a wolf to bay in 1750, although the Welsh naturalist Thomas Pennant considered them to be extinct throughout Britain by 1769. Shortly thereafter, a supposed wolf was killed in Teagues Valley, Carmarthenshire, while another story from Snowdonia tells of a wolf hunt on its northern slopes in 1785 following a spate of attacks on sheep.[17] Though one version of this account relates that it was two brothers who found a den with cubs in it and killed them, Arthur Thomas who farmed near the site had a different story. His grandfather told him that that a soldier had shot the wolf in a fold in the landscape between the Lower and Upper parts of the mountain and that both the mother wolf and her pups had been buried at a location nearby.[18]

On Dartmoor, the last two wolves were believed to have been finished off at Brimpts or Rustington around 1780. The Reverend Weldon Peek wrote that as a boy in 1894 he was told by George Counter, then aged sixty, that his grandfather (in his nineties when Counter was a boy) used to tell him about wolves in the woods around Rustington when he was a young man and how people were 'afeard of un'.[19]

In Ireland, where their destruction had only been seriously targeted for a comparatively short period of time, they lingered much longer. Although the last verified kill was in County Mayo in 1745, in 1786 a wolf that had been killing sheep was brought to bay by the hounds of John Watson of Ballydarton in County Carlow.[20] Rather a fine oil painting of the old man exists in the archive of the Royal Dublin Society. In the somewhat stark surrounds of his flagstoned kitchen, he sits firm-faced in a red hunting coat at a table. His black polished boots have a turned-down rim of soft tan leather and on the knee of his white breeches a pale Kerry foxhound rests its head. Posed with one leg forward while the other is drawn back, both of his feet are nevertheless planted firmly on the skin of his wolf lying prostrate on the floor.

Rumours of another in the Knockmealdown Mountains of Kerry in 1770 or the Wicklows are impossible to prove and, although it's just possible that they may have survived in the Sperrins until 1810, a report of a wolf being killed on the Dingle Peninsula in 1829 is likewise no more than rumour.[21]

But no explanation has ever been offered for a wolf killed on Cader Idris in 1882, which was believed to be an escape from captivity. Nor for another that, while never brought to justice, attracted the headline of 'A Supposed Wolf in Wales' in 1886 in *The Field* magazine. Whatever it was, the readers learnt: 'In the neighbourhood of Lanover park, an animal supposed to be a wolf, has devoured forty lambs and several hundred geese, ducks, chickens and pheasants.'

On 16 May 1916, in Penmachno in Machno Valley, the newspaper *Adsain* reported that 'Farmer Battles with Wolf in an Exciting Encounter in the Welsh Mountains'. The article noted that:

Last Monday a farmer from Cwm Penmachno was shepherding his flock when he saw a large animal lurking aside a large stone tower ... After moving to a more advantageous place the farmer fired at the beast and injured him so badly that he hopped in the direction of a narrow gorge where he stood threateningly. As the man neared it the animal leapt at him ... and was shot dead. After examining the body, the farmer was convinced that he was a huge wolf and that was the opinion of others who saw the body.[22]

This is a very late date for a story that at the time excited no attention at all. Amid the lengthening lists of the dead from the First World War, *Adsain*'s report was never followed up and, although it was supposed that the 'wolf', if indeed that's what it was, had escaped from a travelling circus, no evidence of any sort for this supposition was ever advanced.

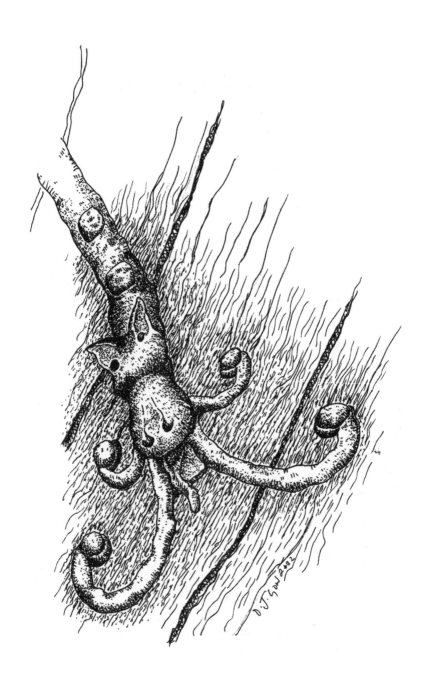

CHAPTER TEN

A Celestial Hammer

In memory of the wolves,
part of these lands,
lost to generations.
we await your return.

Beatrice Searle, *Foregathered wi' the Beast*

One day the iron wolf found me.

I was happily book shopping in Hay on Wye. Although I normally never bother with the boxes outside on the small tables where pamphlets and guidebooks are stacked, a small faded booklet in blue with a black-and-white photo on its front of a plump carved face, all beard, boughs and curled leaf-hair, a carved stone image of a medieval green man, caught my eye. It was titled *A Short Account of the Church of Abbey Dore*. When opened, its central pages revealed other images of the church and, clearly beneath one, were the words: 'wolfs head on hinge'. Excitedly scanning the full text, I identified on its last page that the head was located on an internal north doorway believed to date back to the thirteenth century.

Had I looked well enough outside but not in?

The booklet produced by the church itself addition- ally informed the reader that the wolf was 'believed to commemorate the order given by Edward I that a special effort was to be made in this and neighbouring counties to exterminate wolves'.[1]

Not any Welsh tribute from Edgar centuries before, as Russell Coope had claimed, but rather an instruction to a marcher lord, Sir Peter Corbet, handed down on 14 May 1281. The 'Marches', as the borders between England and Wales were termed, formed a frontier. Sep- arated by geography and jurisdiction, the lords that ruled there had privileges that set them apart. Royal writ did not work and they ruled their own lands in a manner that Gilbert the Earl of Gloucester stated to be *'sicut regale'* – 'like kings'. They built castles, administered laws, waged war, established markets in towns, appointed their own deputies and sat their own sheriffs in their own courts at whim.

In short, they had full jurisdiction over all matters of all sorts except high treason.

They maintained their own records, nearly all of which are lost and, as a result, no account in detail exists of Corbet's activities, although King Edward I was clear in intent:

Ye are to know that we have enjoined our beloved
and trusted Peter Corbet to capture and destroy
wolves, wherever they may be found in all of the
forests, parks and other places within our counties
of Gloucester, Worcester, Hereford, Shropshire and
Stafford, bey every means and his own native cunning, and for that reason we command you that you
assist him with the same end in view.[2]

The only other mention thereafter of this affair occurs
on 29 April 1300, when his huntsmen were paid 6s and
8d to bring Peter's dogs to the king on his death.

While his wolf dogs and their descendants would be
required for centuries thereafter, the end of the wolf
finished the dogs that were bred to destroy them. The
modern Irish Wolfhound is a nineteenth-century cross
between deer hounds and Great Danes. As such, it's a
gentle soul well removed from the savagery of the seven
hounds exported from Ireland in 393 CE to the circus
of Rome, where they excited great attention on account
of their strength and fierceness. Desired by the Indian
emperor Jahangir, they were also much admired by the
Persian rulers, who recorded that 'in Ireland there are
hounds ... of great strength and size and fine shape'.[3]
Centuries later, Bishop Lesley described their Scottish
equivalents as 'being larger than a year-old calf',[4] and
at least one Scottish laird, called Robert Grier of Lag, in
Dumfriesshire, used them to hunt and kill 'covenantors'

protesting against the imposition of the Stuart kings in the 1600s when he was not otherwise hunting wolves.[5]

In Ireland, these large, fierce 'Wolfe dogges' were so valuable that in 1652 Cromwell issued an order of declaration against their export as they were so badly needed. However, by 1874, when Thomas Bell wrote his *A History of British Quadrupeds*, only eight could be found, in the care of Lord Altamont. Although Bell considered these to be degenerate in comparison to the race as it once had been, with 'the tip of the nose to the end of the tail, five feet one inch' and 'from the toes to the top of the shoulder, two feet four inches and a half', they remained formidable till their end.

———

It was time to return to Abbey Dore and in the late afternoon I drove for less than an hour to the old church. While the birds sang still in the trees the wild flowers outside had faded by late summer. No children this time. No bull of brick red.

By chance, as I walked down towards its entrance, a pleasant man perched on a fallen carved column turned out to be a voluntary warden ending an afternoon of groundswork. When he enquired of my purpose and I told him the wolf tale, he stated that he knew nothing of it but was clear that an old door masked from the outside by its new external partner was still in place in the north wall next to the carved oak pew.

Yes, he knew where the key was and yes, he was willing to help. Only our footsteps broke the silence as we walked the worn slabs towards the door in its arch of carved stone. The heavy iron key creaked and clicked in the lock when turned by his hand. We could see the top hinge quite clearly. Long with two sets of whorls like the extended tips of curled wood ferns. No wolf head was there.

At the base it was dim. Separated from the outer door by only a few inches, a detritus of mortar caught by dense spiders' webs formed its shroud. As I knelt to uncover what lay below, a dead mouse dried and hairless fell free.

There was a shape underneath. Further wiping with a hurriedly obtained rag exposed by mobile phone light the crude form of a head. Short ears laid back at either side of a broad-backed skull. A raised forehead and temple. A long blunt-tipped snout with a nose protruding slight but clear.

The iron wolf was there.

Created by a master who, like the stone carvers of Kilpeck church only eight miles away, knew its form. Did he observe a live beast, captive or bound? He could have drawn from a severed head nailed to a tree, beam or church door by Sir Peter after a sweated day's chase. A wrought-iron nail through its neck skin or a spike bashed back into wood between the void of its glazing eyes.

We looked for a time at his handiwork.

A last link quite tangible. Not to the flattened remnants of a dying race but a testament to their being once vital and whole.

When the warden eventually said he must leave as his dinner was waiting, I thanked him for his help. We pushed the door shut, turned the key and walked away, leaving the iron wolf alone once again in the dark.

———

The young wolf crumpled and fell as if hit by a celestial hammer.

Videoed on the mobile phones of enthralled Danish watchers, she had been idly, curiously, following a tractor as the farmer ploughed. Disturbed by the arrival of a black pickup truck, she looked up and then turned to move back towards the trees. The impact of the bullet, which sped with a cluck from the barrel of the rifle's silencer, drove her body tail-flailing, face-first into the soil. The old man who pulled the trigger in response to his son's summons delayed not at all when he realised what the observers had seen. By the time they stood up to shout, he was driving away.

A few years ago, I went to visit a friend of mine in the Danish Forest Service to view the activities of a beaver population that the Ministry of Environment had released in the forest of Klosterheden on the Jutland peninsula. As we sat in a waiting room next to his office one morning, drinking strong black coffee from small white cups, Thomas waved to us as he talked on his phone and mimed with extended fingers that he would be ten minutes. His call took much longer and, when he came to meet us, head shaking, he explained that it had been a conversation with a school teacher who, distraught by a recent news announcement that up to eight wolves were believed to be free living in Denmark, had begged him to instruct his rangers to kill them all. Convinced that her tiny wards were going to be eaten in the grey of an early morning as they waited for the school bus or under the orange glow of street lamps as they walked home in laughing packs through the snow, she was not at all soothed when he told her that wolves were legally protected.

Almost a decade on and, although to date no children have been consumed, concerns of the kind expressed by the teacher are glacially slow to change. Thomas's observation that 'the wolf divides the water' is internationally apt. Every Dane has an opinion about the return of the species. Hate and fear on the tip of one rainbow's end, admiration and longing at the other. While retired dentists or doctors who buy small farms in the countryside on which to keep herds of rare cattle or sheep are generally incensed when the remains of one of their fondling is spread around its paddock in a 30-foot arc, it's not exactly significant damage.

On Christmas Eve 2019, in what was probably a world first, a pet kangaroo was 'eaten as Christmas dinner' by a wolf in a pen near Balen in northeast Belgium.[6] While its mate was bitten on the ear, it survived and, though no doubt shocked to its hopping core, went on to make a sound recovery.

Incidents of this sort are certain to rise as we keep more of what were once farm animals as novelty pets and other more exotic fancies. Wolves have no ability to distinguish the creatures we love from the ones we don't really and, although it's annoying for the owners to pick up the half-eaten collars or examine the tooth dents in the bells that their cherished once wore, theirs is a personal loss.

But events of this sort ensure that wolves remain highly political and when, on 1 September 2022, wolf GW950m killed and partially ate a pony called Dolly in a small village in Germany, he squarely hit the mother lode of dissent. Unlike any of the twelve other previous victims that his DNA had linked him to, Dolly was a pony of a different sort.

Her owner was the president of the European Commission, Ursula von der Leyen.[7]

In the weeks following Dolly's death, Ursula ordered her officials to carry out an in-depth analysis into the dangers posed by wolves after she developed a much more focused interest in this Europe-wide phenomenon. Although GW950m, it transpired, had been sentenced to death for his criminal activities the day before Dolly died, von der Leyen's bereavement attracted considerable international media attention. Officials in Hannover placed a bounty on his head and, despite a legal appeal against his execution from the Society for the Protection of Wolves, the court upheld their official position on GW950m, who had by then been renamed 'Snowy' by his wolf protectionist pals.

While existing laws that protect wolves within the EU allow for those that specifically predate livestock to be shot, this dispensation is not random and does not extend to any other unimplicated pack members. If this all seems rational, then Pekka Pesonen, the secretary general of agricultural cooperatives in the EU, who was swift to agree with the grieving commissioner, provided an alternative view. As soon as the media were listening, he opined that wolf attacks were forcing farmers in France, Germany, Spain, the Baltic states and Finland to quit their profession and that, 'Wolf populations needed to be managed ... If the commission doesn't recognize this, the consequence will be frustration (that) will boil over in the rural communities.'[8]

From a hi-tech agribusiness viewpoint, this is nonsense as the producers of carrots, kumquats or melons in high-rise hothouses will be utterly unaffected by wolves. Although more than a few of his members who are capable of rational thought will otherwise be unfazed, the Danish farmer who shot the wolf did so to protect his threatened carrots. If you substitute beaver, goose, pine marten, eagle or badger for wolf in Pekka's prepreparedtext for farming fools then you have a standard agricultural response from a mindset that's medieval.

In the Scandinavian countries of Norway and Sweden where, in theory at least, the same laws that apply to the protection of wolves in the EU should also hold sway, a death-cult based around views of that sort is already entrenched. In the bowels of the Swedish Parliament in

Stockholm is a shooting gallery where representatives of the Swedish Hunters' Association lobbied in 2023 to reduce the national wolf population from around 460 to 170 individuals.[9]

Although an independent scientific report stressed that the population was physically fragmented and genetically fragile, wolves in Sweden luckily remain only severely endangered. In neighbouring Norway, where their status is critical, the government, without any rational reason, sanctioned the killing of 60 per cent of the population in 2022 to ensure their wolf population remains poised on a knife edge of extinction.[10]

Physically present but diluted and destroyed as a life force.

Although this plan was dropped after campaigners secured a court injunction, Siri Martinsen, the chief executive of the animal rights organisation Noah, which campaigned against the cull, stated the government simply 'don't want wolves in Norway, they have always been clear that wolves should not be breeding in nature'.[11]

The wolf killed in Denmark was on a farm belonging to Steffen Troldtoft, a former parliamentary candidate with the Danish Liberal Alliance party. His 66-year-old father was arrested and charged with violations of Denmark's hunting act. His firearm and vehicle were seized and, although he initially denied killing the wolf, he was charged and brought to trial in 2018. The state prosecution service pressed for a custodial sentence of six months of unconditional imprisonment, but by then his health was failing and the final verdict, when it came, was forty days' probation and a suspension of his hunting licence for two years.

The film of his actions generated significant international attention and considerable national debate. The day after its broadcast, one reporter visited a supermarket in Ulfborg to interview locals, who were clear in their views that the wolves had to go. In their opinion, the wolf killer was a hero and any wolves that remained should be captured, loaded onto a lorry, driven to Copenhagen and released there into the public parks.

Field sign observation coupled with DNA analysis of their scats identified thereafter that what was left of the pack split up of their own accord, with most of the remaining individuals travelling back along the costal routes that they or their parents had followed from northern Germany. There, a sister born in Denmark found a mate and, in his company, returned home. This pair bred and established a pack with five pups in 2019. Although she was observed to be pregnant in 2020, no cubs were ever seen and, later that season, her mate disappeared. Vanishing wolves that are illegally killed are, it transpires, as common a phenomenon in Denmark as they are elsewhere in the wider European Union: ecologists identify from the DNA in wolves' scats that they have come together to breed, then they disappear without a trace very shortly thereafter.[12]

But the recovery of the wolf in Western Europe is without doubt a conservation success story. In September

2022, Rewilding Europe in their *Wildlife Comeback in Europe Report* suggested that nearly seventeen thousand wolves have returned to reoccupy landscapes in twenty-eight countries from which they were cleansed many centuries ago.

New views, as a result of their presence, are rising in result. When Laura Serrano Isla, who owns a flock of 650 sheep near Burgos in northwest Spain, was interviewed in 2021 she observed that, 'There have always been wolves. We humans have hunted and killed all the animals around us because we want everything for ourselves. We think we rule the world but if we kill all the rest of the animals, the wolf will come for our livestock.'[13] Although Laura's views are not popular among her fellows, her voice would once have been hardly heard.

At a recent European beaver meeting, I asked several biologists about how the wolves in their landscapes were faring and received a mixed bag of response. In Southern and Eastern Europe, where attitudes in countries like Greece, Croatia and Romania have long been relaxed, they are protected and their populations are strong. Although commonly scapegoated by the sensationalist press and unscrupulous politicians for the danger they present to people, or for killing sheep as they always have done, most people are unconcerned.

The slowness of the official compensation schemes for any livestock they do kill is the biggest issue of tension. In Luxembourg, where they have just arrived and live-stock compensation schemes are working well, there is no issue. In Germany, where nearly five hundred cubs are now born in established packs annually, there is both acceptance and illegal killing, even though sheep keeping is declining rapidly and solutions such as electric corrals, guard dogs, exclusion fences and compensation are all well in place. In Belgium it is the same, with some NGOs providing wolf-proof fencing free of charge.

This assembly of sensible solutions masks the crude reality that, with technology on our side, we know very well that less than nothing remains to fear from the wolf. As an extreme, if you wanted to restore wolves in Britain but were of a geeky, control freak sort of mindset, then every released individual could be fitted with a radio-satellite transmitter in a neck collar to guarantee they could always be found. Given that this would appeal very greatly to modern conservationists who like to remain firmly in charge of nature, in locations like British Columbia it's also currently being used as a technique for wolf management. The system works on the basis that individual wolves, which are captured alive in leg hold traps and fitted with a satellite transmitter, are used to lead government hunters back to their uncollared packs, which are then swiftly eliminated by marksmen from helicopters. The collared or 'Judas wolf', which is left alive, will then search in desperation for a new mate or pack, affording the hunters the ability to locate and kill many more wolves.[14]

Would this well-proven option satisfy both the inept and the murderous?

A decade ago, on a field trip to a forest near Passau, my bearded, bohemian pal Chris Price, who now leads the sheep-loving Rare Breeds Survival Trust, asked our group leader Gerhard if wolves were ever there. Gerhard's negative response was contradicted by the forest's owner, who said that they did come but always died.

Poisons, snares or bullets were their welcome from the local hunters whose age-old saying of '*schieBen, shaufeln, schweigen*' – 'shoot, shovel and shut up' – are the same words used in Britain by the killers of raptorial birds. When later in the week, Chris enquired of Thomas Obster, the tall, neat, genial representative of the Bavarian Farmers' Association, who has kindly met us for decades in his sparklingly clean Mercedes to unconvincingly explain why beavers have reduced his members to a level of near medieval beggary, he replied it was because wolves were officially 'unacceptable' for his organisation. When asked why again, Thomas said it was because they were a big problem for sheep, and when Chris, who can be determinedly persistent, pointed out that having driven around Munich for nigh on a week we had hardly seen a sheep, Thomas smiled and told him that it was 'not that the wolves that were there were eating the sheep that were not there, but rather that if the sheep that were not there, were there, they almost certainly would'.

In the wake of this masterful political response, Thomas did say with a chuckle, off the record until now at least, that, while the wolves were really good at making the roe deer that fed in the shelter of the maize fields very nervous about doing so, no potential positive of this sort could ever be aired by the Farmers Union as they were politically allied to the German Hunting Association, whose members, wishing to eat deer themselves, were staunchly anti-wolf.

In Britain, the clunking melodies of the sheep bells and their hooded human followers have faded far beyond memory. No collar to protect the neck of a great hound with its sharp steel spikes set into a band of blood-stained leather or brass-hafted hunting horn remains that can be linked directly to the death of a British wolf.

The bounties are paid. The pits are infilled.

Could and should there be a tomorrow?

The English Channel prevents the ready return of the wolf and leaves those of us now living in Britain dispossessed.

No chance to recant or recall.

In an isolated scallop in the Galloway glens tucked well into a landscape very near to the English border lies Lag Maddy Chriach or the 'hollow of the plundering wolves'. In Denmark, far right politicians demanding that 'burqas and wolves' should be banished have promised the strengthening of the Danish–German border in order to secure property rights against gangs of Eastern European thieves and 'plundering' wolves.[15]

If we bring wolves back, they will for sure plunder once more. Not too much to begin with but in the end, as their numbers increase, in plenty.

This truth is simple and, despite the solutions developed by other nations, their prospects on our island will forever be confounded by their old enemy – the sheep.

Sheep are not evil; it's how we keep them that hurts. Without vast public subsidies, they would simply cease to exist and yet, any time a proposal to reintroduce a species as modest as a woodcat or marten is proposed in extreme rural circles, the ridiculous prospect that they might 'eat a lamb' seals their doom.

Pipsqueak views of this sort dictate near utterly the future of all other life on our island.

It is possible, however, that another ungulate that wolves will readily consume could hasten a case for its return. Although studies in Europe indicate that wolves currently have little impact on deer numbers, this is a phase that will pass as their numbers increase. In areas like the Veluwe in the Netherlands, even a low wolf density has forced wary prey species to alter their movement patterns to avoid them and, by doing so, enabled long-suppressed tree seedlings and wild flowers to regenerate.

Might a time of edgy movement, of caution, fear and alert come fast to Britain's forests where the six species of deer – red, roe, fallow, sika, muntjac and Chinese water deer – that dwell within them will be forced to change their ways? Would the boar population, expanding expeditiously at present, be brought down to earth with a bump? As our national deer herd of perhaps two million continues to expand by around 10 per cent annually without any real checks, could the wolf resume its ancient mantle of forest guardian?

All of this is possible and has indeed been foreseen. Nearly a century ago, a mild American naturalist called Aldo Leopold predicted its truth. Originally employed in the early twentieth century as a state trapper of predators to protect game and livestock, his chance encounter with a female wolf he had shot reformed his understanding forever.

> We reached the old wolf in time to watch a fierce green fire dying in her eyes. I realized then ... that there was something new to me in those eyes – something known only to her and to the mountain ... I thought that because fewer wolves meant more deer, that no wolves would mean hunters' paradise ... Such a mountain looks as if someone had given God a new pruning shears, and forbidden Him all other exercise. I now suspect that just as a deer herd lives in mortal fear of its wolves, so does a mountain live in mortal fear of its deer.[16]

When in 2020 the Department for the Environment Farming and Rural Affairs announced to the amazement of anyone credulous enough to believe it that they were going to boldly set up a task force to restore lost species such as the wolf and lynx, a baying pack of farmers harried the then minister Zac Goldsmith to apologise and chased his underlings back into their burrows.[17] A year or so on, when new incumbent Thérèse Coffey confirmed to the politburo of the National Farmers Union that she would not support his stance, they lapped up her words like barnyard cats with a bowl of chicken grease.[18]

In the end, politicians will do what they feel to be right – for them. As rural power blocks dwindle, a growing community who understand that Britain is one of the most nature-impoverished nations on the planet, and that doing nothing is simply not an option, want action. Many well-informed films and documentaries produced independently by young people, research articles in scientific journals and sympathetic news media reports of all sorts have in recent years presented a well-reasoned case for wolf restoration. One scientific study undertaken recently indicated that between 10,139 km^2 and 18,857 km^2 of Highland Scotland would afford suitable wolf habitat and further estimated that this would be sufficient to support between 50 and 94 packs.[19]

In terms of habitat and prey availability, wolf reintroduction is feasible. It's the cultural fear of the dark in which we've swaddled ourselves that will be hard to overcome.

Even so, their appeal is rising.

In 2015, nearly nine thousand individuals, young and old, enjoyed a wolf hunt in Bury St Edmunds. There, they pursued life-size models of wolves created by eighteen local artists. Local dignitaries attended and business organisations supported the event to entice more tourists, showcase local craftsmen and make the town's residents more generally aware of their historic links to the legend of St Edmund. In the same year, *The Wolf Border*, a novel by Sarah Hall, was published, about the owner of a great estate in the North of England who releases his own study group of wolves from the bounds of their enclosure. As the wolves wander back into their old haunts, refuge is offered to them by a socialist Scotland with an executive and society well sick of the politics of their southern overlords and they are allowed to remain.

Wolves are not the bad guys any more.

Aldo Leopold died in 1948, forty-seven years before the release of twenty-one wolves captured in Canada realised his vision of restoring them to the Yellowstone National Park. Although this project remains caustically controversial, one study undertaken in its early years identified that wolf watching alone in the national park was generating 35 million dollars in direct revenue by 2006, with indirect spends in the same year amounting to an additional 70 million.[20] In Spain, a similar embryonic project in 2013 was reported in the *Financial Times* to have generated 176,000 euros in a single season.[21] Wolf-watching tours are available now in France, Italy and Germany, where their spoor, spraint and prey remains can be found in the company of expert guides. One day soon when the farm subsidies finish, other income streams will have to be sought by those who own land.

In the far north of Britain lies the 4,000-km² wilderness of the Flow Country. The cartographer Timothy Pont, having observed in 1596 the 'extreem wilderness' of the landscape with its dark watered peat pools and prairies of billowing bog cotton, also recorded in pencil on his map that it contained 'many woolfs'. He noted the sport that their hunting provided and inferred that the landscapes surrounding the River Naver in its centre – Strathnaver – 'never lack wolves more than ar expedient'.[22] Did this comment suggest that those that endured in his time were tolerated? Nearly fifty years later, Sir Robert Gordon again referred to them as a desirable quarry and considered that 'driven from almost all the rest of the island, they seem to have fixed their lairs and their homes here [in Strathnaver]. Assuredly they are nowhere so plentiful.'[23]

I believe that the last wolves in mainland Britain lived and died in this sanctuary, protected by its complicated vastness and, for a time, by the hunters who cherished their pursuit. Nowhere else do writers speak of their being plentiful at the end. Clusters of folk tales regarding the splay of late killings that occurred on the Findhorn, Shenval and Brora all make sense if this was their final bastion. While no hard evidence exists to back this conjecture, perhaps the 1st Duke of Sutherland well knew of their end. In stone, still he stands resolute on the summit of Beinn a' Bhragaidh, at the tip of his 76-foot-high pedestal, the Mannie. Although he died in 1833, it was he who brought the sheep from the hills of the Cheviots to graze his pastures ever after once his henchmen had forced his clansmen overseas. On his coat of arms, two wolves with collars and chains rise on their hind legs to clasp firmly the family crest. While there is no record of his killing or enslaving any of their kind, the sheep that came north brought with them the most determined of border shepherds long baptised in the fonts of deep bigotry. These religiously unwavering believers in the sanctimony of sheep would without doubt doom the very last of the wolves.

A final story remains, from the *Northern Times* on 26 September 1929. In 1888, the guest of a shooting tenant caught in the late afternoon of a day in a thick hill fog decided as night fell to stay in a cave in the valley of the River Dionard. After making a fire to warm himself he fell asleep and awoke to:

A pair of sunken baleful looking eyes regarding me steadily and stealthily across the dying embers of the fire. I slipped a couple of cartridges into my gun, and as I did so I heard a low painful whine. I could now make out a white form like a huge dog lying not more than three feet from me. Its head rested on its paws ... the face showed signs of great age. I stood up with my gun at my shoulder, but the beast did not move, and I could not find it in my heart to shoot ... The creature then rose, and I saw to my unbounded astonishment that a great silver-grey female wolf faced me ... in her eyes brooded a look of unutterable loneliness and misery.[24]

If it is an honest account, the Dionard is well within very easy loping distance for a wolf from Strathnaver.

So, as this new century begins, wolves are both return-ing themselves and, more astoundingly, being brought back by people to where they should be. Though individ-uals, communities and societies have risen to greet them and celebrate their return, hatred is a virulent pox and many more wolves will still meet their ends at our hands. But though great cruelty is without doubt still to come, we are now on course worldwide to forge an entirely new relationship with the wolf.

One day, when wolves are brought back to Britain, their return will not simply be about land healing. We will consciously or not be healing ourselves. Atoning for the great wrongs we have inflicted on this incredible creature, appeasing the loss of a life-filled landscape that we alone have reduced to dereliction.

When I had time or had work to do in their enclosure, such as building elevated viewing platforms, which they enjoyed very much for the purpose of generally observ-ing the happenings of the park or, better still, as a launch pad for bouncing on people, I played with my cubs. I am five feet eight inches in height and stocky, and, although Nadia's landings generally knocked me over, they winded any smaller beings that she chose to use as trampolines.

I left my employment there in 2003 and had to leave them behind. For a time in the years that followed, I was not allowed to return, but, one Sunday years after when no one was watching, I paid my gate ticket and snuck in to see Nadia. Mishka had died a year or so before but Nadia, old and tired, was sleeping half in and half out of her shed. When I called her, she awoke and unsteadily on shaking legs she came towards me to have her chin stroked through the wire.

In youth, her amber eyes had twinkled with the brightest of stars and I realise now that both she and her kind are forever impressed in the landscapes of Britain. In pattern and spirit, they are the reds of the eastern sandstones, the blue of the fenland clays, the umber and russet of the autumn's turning leaves, the burgundy of the heather. If the wind blows softly through the tall wheat fawn of the moor grasses the wolf is moving through and, when the storms skirl through the gun-metal greys of the winter skies in rising crescendo, their fierce calls are and always will be forever eternal.

— ACKNOWLEDGEMENTS —

John Smellie, Kent Woodruff, Terry Coult, Ben Coult, Peter Cooper, Dr Danielle Schreve, Jennifer Gallichan, Nick Lindsay, Carol Davies, James Hall, Ian Brodie, Dr Lee Ray, Roy Dennis, Cardiff Museum, Wally Jones, Thomas Borup Svendsen, Gary Easton, Robert Coope, Sarah Cross, Julia Noble. The wolves I have known and the brilliant Twitter folk who provided a wealth of odd and erratic content.

— NOTES —

Prologue: A Last Wolf

1   Andrew C. Kitchener, 'Extinctions, Introductions and Colonisations of Scottish Mammals and Birds Since the Last Ice Age' in *Species History in Scotland*, ed. Robert A. Lambert (Edinburgh: Scottish Cultural Press, 1998).

2   John Sobieski and Charles Edward Stuart, *Lays of the Deer Forest* (Edinburgh: William Blackwood and Sons, 1848), 2:245–7.

3   Thomas Dick Lauder, *An Account of the Great Floods of August 1829, in the Province of Moray, and Adjoining Districts* (Edinburgh: Adam Black, 1830), 45–6.

4   David Stephen, quoted in Jim Crumley, *The Last Wolf* (Edinburgh: Birlinn, 2010), chap. 6, ebook.

5   Jim Crumley, *The Last Wolf* (Edinburgh: Birlinn, 2010).

6   John Gordon, ed., 'Duthil, Elgin' in *The New Statistical Account of Scotland* (Edinburgh: William Blackwood and Sons, 1845), 13:126–7, distributed by the University of Edinburgh and University of Glasgow, 'The Statistical Accounts of Scotland Online Service' (1999), https://stataccscot.edina.ac.uk:443/link/nsa-vol13-p127-parish-elgin-duthil.

Introduction: Hunting the Wolf

1   Cledwyn Fychan, *Galwad Y Blaidd* (Aberystwyth: Y Lolfa, 2006).

2   Ilse Köhler-Rollefson, *Hoofprints on the Land* (London: Chelsea Green Publishing UK, 2023).

3   Michael Wood, *The Story of England* (London: Penguin, 2010).

4   Andrew E.M. Wiseman, '"A Noxious Pack": Historical, Literary and Folklore Traditions of the Wolf (Canis lupus) in the Scottish Highlands', *Scottish Gaelic Studies* 25 (2009): 118, https://electricscotland.com/nature/A_Nox ious_Pack_Historical_Literary_and.pdf.

5   New Forest Commoner, 'New Forest: Forest Laws, Punishment and Reform', 24 August 2021, http://newforestcommoner.co.uk/2021/08/24/ new-forest-forest-laws-poaching-punishment-and-reform.

6   'Extracts from Liberties in the County of Dorset', Andy Poore, personal communication to the author, 2021.

7   William Scrope, *Days of Deer-Stalking* (Glasgow: Thomas D. Morison, 1883), 285–6.

8   Timothy Pont, *Glasgow and the County of Lanark – Pont 34*, 'National Library of Scotland', ca. 1583–1596, https://maps.nls.uk/view/00002331.

9   William Forrest, *The County of Lanark from Actual Survey*, 'National Library of Scotland', Edinburgh: n.p., 1816, https://maps.nls.uk/view/74400274.

10   David Prentice Menzies, *The 'Red and White' Book of Menzies: The History of Clan Menzies and its Chiefs* (Glasgow: Printed for the Author by Banks & Co., 1894), 108, https://digital.nls.uk/histories-of-scottish-families/ archive/96654652.

Notes

Chapter One: The Iron Wolf

1   Cledwyn Fychan, *Galwad Y Blaidd* (Aberystwyth: Y Lolfa, 2006).

2   Adrian M. Lister, 'Late-glacial Mammoth Skeletons (*Mammuthus primigenius*) from Condover (Shropshire, UK): Anatomy, Pathology, Taphonomy and Chronological Significance', *Geological Journal* 44, no. 4 (July 2009): 375–500, https://doi.org/10.1002/gj.1162.

3   Anthony Dent, *Lost Beasts of Britain* (London: George G. Harrap & Co., 1974), 119.

4   Andrew E.M. Wiseman, '"A Noxious Pack": Historical, Literary and Folklore Traditions of the Wolf (*Canis lupus*) in the Scottish Highlands', *Scottish Gaelic Studies* 25 (2009): 96, https://electricscotland.com/nature/A_Noxious_Pack_Historical_Literary_and.pdf.

5   Edward Turner, 'Ashdown Forest: Or as it was Sometimes Called, Lancaster Great Park', *Sussex Archaeological Collections* 14 (1862): 35–36.

6   John Pollard, *Wolves and Werewolves* (London: Robert Hale, 1964), 78.

7   Henry James Morehouse, *The History and Topography of the Parish of Kirkburton and of the Graveship of Holme* (Huddersfield: Printed for the Author by H. Roebuck, 1861).

8   Anthony Dent, *Lost Beasts of Britain*.

9   Derek Yalden, *The History of British Mammals* (London: T. & A.D. Poyser, 1999).

10 Alexander Robert Forbes, *Gaelic Names of Beasts (Mammalia), Birds, Fishes, Insects, Reptiles, etc.* (Edinburgh: Oliver and Boyd, 1905), 227.

11 Gaelic Society of Inverness, *Transactions of the Gaelic Society of Inverness* (Inverness: Northern Chronicle, 1890) 15:292.

12   Sarah J. Wager, 'The Hays of Medieval England: A Reappraisal', *Agricultural History Review* 65, no. 2 (December 2017): 185.

13   Fychan, *Galwad Y Blaidd*.

14   Fychan.

15   James Edmund Harting, *British Animals Extinct within Historic Times* (London: Trübner & Co., 1880), 132.

16   Andy Gaunt, 'Wolves, the Wolf Hunters of Sherwood Forest, Outlaws and the "Wolf's Head"', Mercian Archaeological Services CIC, last modified 24 April 2020, http://www.mercian-as.co.uk/wolfhunter.html.

Chapter Two: 'Here, Here, Here!' Cried the Wolf

1   Musée du Loup, Le Cloitre-Saint-Thegonnec, France, personal communication to author, 2021.

2   Cledwyn Fychan, *Galwad Y Blaidd* (Aberystwyth: Y Lolfa, 2006).

3   Gerard Rennick, 'The "Climate Scientists" Who Cried Wolf'. Senator Gerard Rennick, 26 February 2021, https://gerardrennick.com.au/the-climate-scientists-who-cried-wolf/.

4   Musée du Loup.

5   Charles Dickens, 'Teeth', *All the Year Round*, 22 March 1879, 322.

6   Fychan, *Galwad Y Blaidd*.

7   Musée du Loup.

8   Fychan, *Galwad Y Blaidd*.

9   Rev. W. Thompson, *Sedbergh, Garsdale, and Dent* (Leeds: Richard Jackson, 1892), 265.

10   Aleksander Pluskowski, *Wolves and the Wilderness in the Middle Ages* (Woodbridge: Boydell Press, 2006).

Notes

11 Kieran Hickey, *Wolves in Ireland: A Natural and Cultural History* (Dublin: Four Courts Press, 2013).

12 Anthony Dent, *Lost Beasts of Britain* (London: George G. Harrap & Co., 1974).

13 Hickey, *Wolves in Ireland.*

14 Andrew E.M. Wiseman, "'A Noxious Pack": Historical, Literary and Folklore Traditions of the Wolf *(Canis lupus)* in the Scottish Highlands', *Scottish Gaelic Studies* 25 (2009): 117, https://electricscotland.com/nature/A_Noxious_Pack_Historical_Literary_and.pdf.

15 'Edinburgh Commissary Court: Register of Testaments', 1514–1850, CC8/8/21, National Records of Scotland, Edinburgh.

16 James Edmund Harting, *British Animals Extinct within Historic Times* (London: Trübner & Co., 1880), 169.

17 Musée du Loup.

18 Fychan, *Galwad Y Blaidd.*

19 Alban Butler, *The Lives of the Fathers, Martyrs and Other Principal Saints*, ed. Frederick Charles Husenbeth (London: Henry and Co., 1857), 2:700.

20 Pluskowski, *Wolves and the Wilderness.*

21 Dent, *Lost Beasts of Britain.*

22 Ian Woodward, *The Werewolf Delusion* (New York: Paddington Press, 1979), 21.

23 Barry Lopez, *Of Wolves and Men* (New York: Scribner, 1978), 236.

24 Wiseman, "'A Noxious Pack"', 100.

25 Carl Seaver, 'Dina Sanichar; The Feral Boy Who Inspired the Jungle Book', History Defined, last modified 4 November 2022, https://www.historydefined.net/life-of-dina-sanichar.

26 *Entrelobos (Among Wolves)*, directed by Gerardo Olivares (2010, Madrid: Wanda Visión), film.

27 Kate Cherrell, 'The Hexham Heads', Burials and Beyond, 13 June 2021, https://burialsandbeyond. com/2021/06/13/the-hexham-heads.

28 Charles Christian, 'When I was a Child, I Remember', quoted in James Campbell and Michael Moran, 'Werewolf Hunter Reveals UK's "Paranormal Triangle Where Terrifying Beasts Gather"', *Daily Star*, 30 May 2021, https://www.dailystar.co.uk/news/weird-news/ werewolf-hunter-reveals-uks-paranormal-24215553.

Chapter Three: The Howling Gods

1 Andrew C. Kitchener, 'Extinctions, Introductions and Colonisations of Scottish Mammals and Birds Since the Last Ice Age' in *Species History in Scotland*, ed. Robert A. Lambert (Edinburgh: Scottish Cultural Press, 1998), 75.

2 Catherine D. Hughes, 'Killer Whale Facts!', *National Geographic Kids*, 24 April 2017, https://www.natgeokids. com/uk/discover/animals/sea-life/killer-whale-facts.

3 Sibelle T Vilaça et al., 'Tracing Eastern Wolf Origins from Whole-Genome Data in Context of Extensive Hybridization', *Molecular Biology and Evolution* 40, no.4 (13 April 2023): msad055, https://doi.org/10.1093/molbev/msad055.

4 'Historic Settlement Secures Conservation of Endangered Red Wolves in the Wild', Defenders of Wildlife, 9 August 2023, https://defenders.org/newsroom/historic- settlement-secures-conservation-of-endangered- red-wolves-wild.

5 US Fish and Wildlife Service, 'Mexican Wolf Numbers Soar Past 200', press release, 28 February 2023,

Notes

https://www.fws.gov/press-release/2023-02/
mexican-wolf-numbers-soar-past-200.

6   Elizabeth Pennisi, 'Borrowed Gene Blackens Wolves:
    Interbreeding with Dogs Gave Wolves an Evolutionary
    Advantage', *Science*, 5 February 2009, https://www.
    science.org/content/article/borrowed-gene-blackens-wolves.

7   Henry Harvey, *With Essex in Ireland*, ed. Emily Lawless
    (London: Smith, Elder and Co., 1890), 116.

8   Adam Weymouth, 'Was this the Last Wild Wolf of Britain?',
    Guardian, 21 July 2014, https://www.theguardian.com/
    science/animal-magic/2014/jul/21/last-wolf.

9   Dan Kraker, 'The Secret Lives of Wolves in Voyageurs Nat-
    ional Park: They Fish and Eat Berries', MPR News,
    17 December 2018, https://www.mprnews.org/
    story/2018/12/17/the-secret-lives-of-wolves-at-
    voyageurs-national-park-they-fish-and-eat-berries.

10 Margaret Osborne, 'In Alaska, Hungry Wolves Have
    Started Eating Sea Otters', *Smithsonian Magazine*,
    26 January 2023, https://www.smithsonianmag.com/
    smart-news/in-alaska-hungry-wolves-have-started-eating-
    sea-otters-180981509.

11 Vadim Sidorovich, 'Findings on the Interference Between
    Wolves and Lynxes', *Zoology by Vadim Sidorovich* (blog),
    6 September 2017, https://sidorovich.blog/2017/09/06/
    wolves-and-lynxes.

12 Francesco Maria Angelici and Lorenzo Rossi, 'A New
    Subspecies of Gray Wolf (*Carnivora,* Canidae), Recently
    Extinct, from Sicily, Italy', *Bollettino del Museo Civico di
    Storia Naturale di Verona: Botanica Zoologia* 42 (2018):
    3–10.

13   Alex K.T. Martin, 'In Search of Japan's Lost Wolves', *Japan Times*, 5 July 2021, https://features.japantimes.co.jp/japan-wolf-search-index.

14   Shuichi Matsumura, Yasuo Inoshima and Naotaka Ishiguro, 'Reconstructing the Colonization History of Lost Wolf Lineages by the Analysis of the Mitochondrial Genome', *Molecular Phylogenetics and Evolution* 80 (November 2014): 105–12, https://doi.org/10.1016/j.ympev.2014.08.004.

15   John Knight, 'On the Extinction of the Japanese Wolf', *Asian Folklore Studies* 56, no. 1 (1997): 129–59, https://doi.org/10.2307/1178791.

Chapter Four: There I Saw the Grey Wolf Gaping

1   Edward Second Duke of York, *The Master of Game*, eds. William. A. Baillie-Grohman and Florence Baillie-Grohman (London: Chatto and Windus, 1909), 63.

2   Kieran Hickey, *Wolves in Ireland: A Natural and Cultural History* (Dublin: Four Courts Press, 2013), 42.

3   Cledwyn Fychan, *Galwad Y Blaidd* (Aberystwyth: Y Lolfa, 2006).

4   John Caius, *Of Englishe Dogges* (London: Rychard Johnes, 1576), 22.

5   International Wolf Center, *Are Wolves Dangerous to Humans?* (Ely, MN: International Wolf Center, 2003), https://wolf.org/wp-content/uploads/2013/05/Are-Wolves-Dangerous-to-Humans.pdf.

6   Stuart Erskine, *Braemar* (Edinburgh: Andrew Elliott, 1898), 9.

7   Theodore Roosevelt, *Hunting the Grisly and Other Sketches* (New York, London: G.P. Putnam's Sons, 1893), 213.

Notes

8   Thomas Langdale, *A Topographical Dictionary of Yorkshire* (Northallerton: J. Langdale, 1822), 161.

9   Thomas Morton, *The New English Canaan of Thomas Morton*, ed. Charles Francis Adams (Boston: John Wilson and Son, 1883), 209.

10 John Pollard, *Wolves and Werewolves* (London: Robert Hale, 1964).

11 Carter Niemeyer, *Wolf Land* (Boise, ID: Bottlefly Press, 2016), 202.

12 Molly Quell, 'Dutch Court Shoots Down Plan Using Paintball Guns on Wolves', *AP News*, 30 November 2022, https://apnews.com/article/gun-violence-shootings-netherlands-animals-wolves-4b274ae11fde24a51462fb9ce-fa1fc0b.

13 Thomas Ling and Toby Saunders, 'Top 10: World's Most Dangerous Animals', *BBC Science Focus*, 20 June 2023, https://www.sciencefocus.com/nature/what-animals-kills-the-most-people.

14 John D.C. Linnell, Ekaterina Kovtun and Ive Rouart, 'Wolf Attacks on Humans: An Update for 2002–2020', *NINA Report 1944* (Trondheim: Norwegian Institute for Nature Research, January 2021), 25.

15 Harry Cockburn, '"Swarms" of Wolf-dog Hybrids Sweeping Europe, Study Reveals', *Independent*, 23 May 2019, https://www.independent.co.uk/news/science/wolf-dog-hybrid-europe-interbreeding-a8927716.html.

16 Musée du Loup, Le Cloître-Saint-Thegonnec, France, personal communication to author, 2021.

17 John D.C. Linnell et al., 'The Fear of Wolves: A Review of Wolf Attacks on Humans', *NINA Oppdragsmelding* 731 (Trondheim: Norwegian Institute for Nature Research, January 2002), 19.

18   Edward Second Duke of York, *The Master of Game*, eds. William. A. Baillie-Grohman and Florence Baillie-Grohman (London: Chatto and Windus, 1909), 60.

19   Fychan, *Galwad Y Blaidd*.

20   Fychan.

21   James Edmund Harting, *British Animals Extinct* within *Historic Times* (London: Trübner & Co., 1880), 124.

22   Hickey, *Wolves in Ireland*, 48.

23   Eliza Ann Harris Dick Ogilvy, *The Book of Highland Minstrelsy* (London: Richard Griffin and Co., 1860), 254.

24   Andrew E.M. Wiseman, '"A Noxious Pack": Historical, Literary and Folklore Traditions of the Wolf (*Canis lupus*) in the Scottish Highlands', *Scottish Gaelic Studies* 25 (2009): 117, https://electricscotland.com/nature/A_Noxious_Pack_Historical_Literary_and.pdf.

25   Wiseman, '"A Noxious Pack"', 116.

26   Hickey, *Wolves in Ireland*.

Chapter Five: Lords Do Not Rise at the Crack of Dawn

1   Richard Almond, *Medieval Hunting* (Stroud: Sutton, 2003), 70–72.

2   James Edmund Harting, *British Animals Extinct within Historic Times* (London: Trübner & Co., 1880), 133.

3   Almond, *Medieval Hunting*, 70–72.

4   John Cummins, *The Hound and the Hawk: The Art of Medieval Hunting* (Edison, NJ: Castle Books, 2003), 138.

5   Edward Second Duke of York, *The Master of Game*, eds. William. A. Baillie-Grohman and Florence Baillie-Grohman (London: Chatto and Windus, 1909), 62.

6   Harting, *British Animals Extinct*, 136.

7   Raphael Holinshed, Reyner Wolfe, Richard Stanihurst, William Harrison and Edmund Campion, *Holinshed's*

Notes

Chronicles: The Firste Volume (London: George Bishop, 1577), 344.

8   'Wolseley Hall', Lost Heritage, http://www.lostheritage. org.uk/houses/lh_staffordshire_wolseleyhall.html.

9   Joseph Strutt, *The Sports and Pastimes of the People of England* (London: Methuen and Co., 1903), 12.

10 Harting, *British Animals Extinct*, 149–150.

11 Robert Scott Fittis, *Sports and Pastimes of Scotland* (Paisley: Alexander Gardner, 1891), 39.

12 K.M. Brown et al. eds., 'The Records of the Parliaments of Scotland to 1707', 2007-2023, 1428/3/6, University of St Andrews, http://www.rps.ac.uk/trans/1428/3/6.

13 Edward William Lewis Davies, *Wolf-hunting and Wild Sport in Lower Brittany* (London: Chapman and Hall, 1875).

14 John Cummins, *The Hound and the Hawk* (Edison, NJ: Castle Books, 2003), 62.

15 Richard Sidney Richmond Fitter, *The Ark In Our Midst* (London: Collins, 1959), 77.

16 Anthony Dent, *Lost Beasts of Britain* (London: George G. Harrap & Co., 1974), 121.

17 Stephen J. Bodio, 'Of Colonel Thornton and Inedible Quarry', *Steve Bodio's Querencia* (blog), 21 December 2013, http://stephenbodio.blogspot.com/2013/12/of-colonel-thornton-and-inedible-quarry.html.

18 Eloise Kane, 'Wild is the Wolf', *History Today*, 23 January 2019, https://www.historytoday.com/ miscellanies/wild-wolf.

19 Edward William Lewis Davies, *Wolf-hunting and Wild Sport in Lower Brittany* (London: Chapman and Hall, 1875), 25.

20 Davies, *Wolf-hunting and Wild Sport*, 26.

21 Henry Laver, *The Mammals, Reptiles and Fishes of Essex: A Contribution to the Natural History of the County* (London: Simpkin, Marshall & Co., 1898).

22 Vivien Vergnaud, 'Have Wolves Returned to Brittany After a Century Away?' in 'A Glance around France: Bullet proof vests in the Var, and have wolves returned to Brittany?', The Local, last modified 24 September 2018, https://www.thelocal.fr/20180924/a-glance-around-france-bullet-proof-vests-in-the-var.

23 'Wolf Seen in Brittany for First Time in a Century', *Bird Guides*, 12 May 2022, https://www.birdguides.com/news/wolf-seen-in-brittany-for-first-time-in-a-century.

Chapter Six: The Lord Is My Shepherd

1 Thomas Campbell, 'The Soldier's Dream', McMaster Museum of Art, 10 November 2014, https://museum.mcmaster.ca/the-soldiers-dream.

2 Janet Burton, 'Who Were the Cistercians?', *Monastic Wales*, 23 June 2009, https://www.monasticwales.org/showarticle.php?func=showarticle&articleID=3.

3 Susan Rose, *The Wealth of England: The Medieval Wool Trade and its Political Importance, 1100–1600* (Oxford: Oxbow Books, 2018).

4 Lauren Oldershaw, 'History: 150 Years of One of Colchester's Rail Stations', *Colchester Gazette*, 12 December 2016, https://www.gazette-news.co.uk/news/14960684.history-150-years-of-one-of-colchesters-rail-stations.

5 Adrian House, *Francis of Assis: A Revolutionary Life* (Mahwah, NJ: Paulist Press, 2003).

6 Barry Lopez, *Of Wolves and Men* (New York: Scribner, 1978), 146.

Notes

7   *Jn 10:12* (Berean Standard Bible).

8   Anthony Dent, *Lost Beasts of Britain* (London: George G. Harrap & Co., 1974).

9   Alexander Carmichael, *Carmina Gadelica: Hymns and Incantations* (Edinburgh: Printed for the Author by T. and A. Constable, 1900), 1:178–9, https://deriv.nls.uk/dcn6/7583/75832322.6.pdf.

10   Samuel Harvey, *An Anglo-Saxon Abbot: Ælfric of Eynsham* (Edinburgh: T. and T. Clark, 1912), 185.

11   John Caius, *Of Englishe Dogges* (London: Rychard Johnes, 1576), 23.

12   Iris Combe, *Herding Dogs* (London: Faber and Faber, 1987).

13   Thomas More, *Utopia: His Second and Revised Edition, 1556*, trans. Ralph Robinson, ed. Edward Arber (London: A. Constable and Co., 1895), 40–1.

14   Department for Environment, Food and Rural Affairs, 'Livestock Populations in England at 1 June 2023', National Statistics, updated 15 September 2023, https://www.gov.uk/government/statistics/livestock-populations-in-england/livestock-populations-in-england-at-1-june-2023.

15   Julian Kossoff, 'Killer Sheep Discovered Wolfing Down Bird Chicks', *International Business Times*, 30 May 2018, https://www.ibtimes.com/killer-sheep-discovered-wolfing-down-bird-chicks-2685744.

16   Emma Bedford, 'Consumption Volume of Red Meat in the United Kingdom (UK) in 2017, by Type', Statista, 1 February 2023, https://www.statista.com/statistics/642948/red-meat-consumption-volume-united-kingdom-uk.

Chapter Seven: Most Likely a Dog

1 Terry Coult, 'Notes on the Wolf, Bear and Lynx in Durham and Northumberland', *Northumbrian Naturalist* 79 (2015).

2 Danielle Schreve, personal communication to author, 2020.

3 Carina Phillips, Ian L. Baxter and Marc Nussbaumer, 'The Application of Discriminant Function Analysis to Archaeological Dog Remains as an Aid to the Elucidation of Possible Affinities with Modern Breeds', *Archaeofauna* 18 (October 2009): 51–64, https://revistas.uam.es/archaeofauna/article/view/6591/6992.

4 Harry Johnston, *British Mammals* (London: Hutchinson and Co., 1903).

5 Karen Lloyd, *The Blackbird Diaries* (Salford: Saraband, 2017), chapter 10, ebook.

6 Cledwyn Fychan, *Galwad Y Blaidd* (Aberystwyth: Y Lolfa, 2006).

7 Reginald Innes Pocock, *The Field*, 5 July 1928.

8 Fychan, *Galwad Y Blaidd*.

9 Danielle Schreve, personal communication to author, 2020.

10 Schreve, personal communication, 2020.

Chapter Eight: Severing Their Umbilicals

1 Musée du Loup, Le Cloitre-Saint-Thegonnec, France, personal communication to author, 2021.

2 Anthony Dent, *Lost Beasts of Britain* (London: George G. Harrap & Co., 1974).

3 Barry Lopez, *Of Wolves and Men* (New York: Scribner, 1978), 168.

4 'Exchequer Records: James VI Household Books' 1525–1651, E31/15, National Records of Scotland, Edinburgh.

Notes

5   Theodore Roosevelt, *Hunting the Grisly and Other Sketches* (New York, London: G.P. Putnam's Sons, 1902.

6   John Pollard, *Wolves and Werewolves* (London: Robert Hale, 1964), 157.

7   Henry James Morehouse, *The History and Topography of the Parish of Kirkburton and of the Graveship of Holme* (Huddersfield: Printed for the Author by H. Roebuck, 1861).

8   John Gordon, ed., 'Moy and Dalarossie, County of Inverness' in *The New Statistical Account of Scotland* (Edinburgh: William Blackwood and Sons, 1845), 14:102, distributed by the University of Edinburgh and University of Glasgow, 'The Statistical Accounts of Scotland Online Service' (1999), https://stataccscot.edina.ac.uk:443/link/nsa-vol14-p102-parish-inverness-moy_and_dalarossie.

9   Pollard, *Wolves and Werewolves*.

10   Lopez, *Of Wolves and Men*.

11   Howard L. Blackmore, *Hunting Weapons* (New York: Dover Publications, 1971), 197.

12   Pollard, *Wolves and Werewolves*.

13   James Edmund Harting, *British Animals Extinct within Historic Times* (London: Trübner & Co., 1880), 139.

14   John G. Harrison, 'The Last Wolf in Scotland – Some Historical Evidence', *John's Scottish History Pages* (blog), 15 July 2020, https://www.johnscothist.com/uploads/5/0/2/4/5024620/harrison_last_wolf_v._4.pdf.

15   Kieran Hickey, *Wolves in Ireland: A Natural and Cultural History* (Dublin: Four Courts Press, 2013).

16   John O'Hart, *Irish Pedigrees* (Dublin: James Duffy and Co., 1892), 1:800.

17   Harting, *British Animals Extinct*, 166.

18   Dent, *Lost Beasts of Britain*, 120.

Chapter Nine: *Void of Noisome Beasts*

1   Raphael Holinshed, Reyner Wolfe, Richard Stanihurst, William Harrison and Edmund Campion, *Holinshed's Chronicles: The Firste Volume* (London: George Bishop, 1577).

2   'Catalogue of the Mammalia of Northumberland and Durham', *Transactions of the Tyneside Naturalists Field Club 1863–1864* VI (1864): 117.

3   Kieran Hickey, *Wolves in Ireland: A Natural and Cultural History* (Dublin: Four Courts Press, 2013).

4   John Sobieski and Charles Edward Stuart, *Lays of the Deer Forest* (Edinburgh: William Blackwood and Sons, 1848), 2:243.

5   Richard Almond, *Medieval Hunting* (Stroud: Sutton, 2003).

6   Charles Dickens, 'As the Crow Flies', *All the Year Round*, 12 December 1868, 40.

7   Caroline Grigson, *Menagerie* (Oxford: Oxford University Press, 2016), 7.

8   'Local Courts: Papers of the Campbell Family, Earls of Breadalbane (Breadalbane Muniments)', GD112/17, National Records of Scotland, Edinburgh.

9   Cosmo Nelson Innes, ed., *The Black Book of Taymouth* (Edinburgh: T. Constable, 1885).

10   Lee Raye, *The Atlas of Early Modern Wildlife* (London: Pelagic, 2023), 45.

11   James Edmund Harting, *British Animals Extinct within Historic Times* (London: Trübner & Co., 1880),132–3.

12   Cledwyn Fychan, *Galwad Y Blaidd* (Aberystwyth: Y Lolfa, 2006).

13   Raye, *The Atlas*, 44.

14   John Pollard, *Wolves and Werewolves* (London: Robert Hale, 1964).

15   Ian Murray, *Old Deeside Ways* (Ballater: Lochnagar Publications, 2015).

16 Andrew C. Kitchener, 'Extinctions, Introductions and Colonisations of Scottish Mammals and Birds Since the Last Ice Age' in *Species History in Scotland*, ed. Robert A. Lambert (Edinburgh: Scottish Cultural Press, 1998), 75.

17 Pollard, *Wolves and Werewolves*, 82.

18 Fychan, *Galwad Y Blaidd*.

19 Eric Hemery, *High Dartmoor: Land and People* (London: Robert Hale, 1983), 459.

20 Anthony Dent, *Lost Beasts of Britain* (London: George G. Harrap & Co., 1974).

21 Hickey, *Wolves in Ireland*.

22 Fychan, *Galwad Y Blaidd*.

Chapter Ten: A Celestial Hammer

1 Frederick Charles Morgan, *A Short Account of the Church of Abbey Dore*, 8th ed. (Leominster: Orphans Press, 1979), 7.

2 James Edmund Harting, *British Animals Extinct within Historic Times* (London: Trübner & Co., 1880), 143.

3 Harting, *British Animals Extinct*, 188.

4 Peter Hume Brown, ed., *Scotland Before 1700* (Edinburgh: David Douglas, 1893), 123.

5 Patrick Laurie, *Native: Life in a Vanishing Landscape* (Edinburgh: Birlinn, 2021).

6 'Wolf Snatches Pet Kangaroo from Belgium Home', *BBC News*, 25 December 2019, https://www.bbc.co.uk/news/world-europe-50912766.

7 Chris Jewers, 'Ursula von der Leyens Horse Lies Dead in a Pasture Following Wolf Attack – as she is Forced to Deny Ordering Cull in 'Revenge', *Daily Mail*, updated 5 January 2023, https://www.dailymail.co.uk/news/article-11602741/Ursula-von-der-Leyens-horse-lies-dead-pasture.html.

8   James Crisp, 'Wanted: The Wolf that Killed the EU President's Prized Pony', *National Post*, updated 5 January 2023, https://nationalpost.com/news/world/wanted-the-wolf-that-killed-the-eu-presidents-prized-pony.

9   Helena Horton and Beata Furstenberg, 'Sweden's Biggest Wolf Cull Starts but Campaigners Fight On', *Guardian*, 2 January 2023, https://www.theguardian.com/environment/2023/jan/02/huge-swedish-wolf-hunt-will-be-disastrous-for-species-warn-experts.

10 'Save our Wolves', WWF, 2023, https://www.wwf.no/dyr-og-natur/truede-arter/ulv-i-norge/rettssak-for-ulven/save-our-wolves.

11 Helena Horton, 'Norway's Wolves "Saved for this Year" as Animal Rights Groups Fight Cull', *Guardian*, 31 January 2022, https://www.theguardian.com/world/2022/jan/31/norway-wolves-saved-for-this-year-as-animal-rights-groups-fight-cull.

12 Peter Sunde et al., 'Where Have all the Young Wolves Gone? Traffic and Cryptic Mortality Create a Wolf Population Sink in Denmark and Northernmost Germany', *Conservation Letters* 14, no. 5 (17 May 2021): e12812, https://doi.org/10.1111/conl.12812.

13 Stephen Burgen, 'Spanish Farmers Deeply Split as Ban on Hunting Wolves is Extended' *Guardian*, 6 March 2021, https://www.theguardian.com/world/2021/mar/06/spanish-farmers-deeply-split-as-ban-on-hunting-wolves-is-extended.

14 Elizabeth Claire Alberts, 'Wolf Kept Alive Year After Year So Government Can Kill His Friends', *The Dodo*, 1 March 2016, https://www.thedodo.com/judas-wolf-canada-1634797811.html.

Notes

15   Dorthe Nors, 'They Want a Wolf-free Denmark. Will
      Migrants be Next?', *Guardian,* 16 May 2018.

16   Aldo Leopold, *A Sand County Almanac* (Oxford: Oxford
      University Press, 1949), 130–32.

17   Helena Hotron, 'Lynx and Wolves could return to England as
      rewilding task force set up by PM', *Telegraph*, 6 March 2021,
      https://www.telegraph.co.uk/news/2021/03/06/lynx-
      wolves-could-return-england-rewilding-task-force-set-pm.

18   Secretary of State Thérèse Coffey's address at NFU
      conference. DEFRA. 22 February 2023.

19   Vashti Gwynn and Elias Symeonakis, 'Rule-based habitat
      suitability modelling for the reintroduction of the grey
      wolf (Canis lupus) in Scotland', *National Library of
      Medicine* 17, no.10 (21 October 22), https://doi.
      org/10.1371/journal.pone.0265293.

20   Cassidy Randall, 'A rewilding triumph: wolves help to
      reverse Yellowstone degradation', *Guardian*,
      25 January 2020, https://www.theguardian.com/
      environment/2020/jan/25/
      yellowstone-wolf-project-25th-anniversary.

21   Paul Richardson, 'Wolf-watching in Spain', *Financial
      Times*, 1 March 2013, https://www.ft.com/content/
      c341b948-7c43-11e2-99f0-00144feabdc0

22   Timothy Pont, *Eddrachilles; Northwest Sutherland;
      Loch Shin – Pont 3*, 'National Library of Scotland', ca.
      1560–1614, https://maps.nls.uk/view/00002285.

23   Walter MacFarlane, *Geographical Collections Relating
      to Scotland Made by Walter MacFarlane*, ed. Sir Arthur
      Mitchell (Edinburgh: T. and A. Constable, 1907), 2:454.

24   Grace Banks and Sheena Blackhall, *Scottish Urban Myths
      and Ancient Legends* (Cheltenham: The History Press,
      2015), 170.

— SELECT BIBLIOGRAPHY —

Almond, Richard. *Medieval Hunting*. Stroud: Sutton, 2003.

Crumley, Jim. *The Last Wolf*. Edinburgh: Birlinn, 2010.

Dent, Anthony. *Lost Beasts of Britain*. London: George G. Harrap & Co., 1974.

Fychan, Cledwyn. *Galwad Y Blaidd*. Aberystwyth: Y Lolfa, 2006.

Harting, James Edmund. *British Animals Extinct within Historic Times*. London: Trübner & Co., 1880.

Lopez, Barry. *Of Wolves and Men*. New York: Scribner, 1978.

Mech, L. David and Luigi Boitani, eds. Wolves: *Behavior, Ecology, and Conservation*. London: The University of Chicago Press, 2003.

Mortimer, Ian. *The Time Traveller's Guide to Medieval England*. London: Vintage, 2009.

Bibliography

O'Connor, Roderic. *An Introduction to the Field Sports of France*. London: John Murray, 1846.

Pluskowski, Aleksander. *Wolves and the Wilderness in the Middle Ages*. Woodbridge: Boydell Press, 2006.

Pollard, John. *Wolves and Werewolves*. London: Robert Hale, 1964.

Rose, Susan. *The Wealth of England*. Oxford: Oxbow Books, 2018.

Strutt, Joseph. *The Sports and Pastimes of the People of England*. London: Methuen and Co., 1903.

Tisdall, Michael W. *God's Beasts*. Plymouth: Charlesfort Press, 1998.

Yalden, Derek. *The History of British Mammals*. London: T. & A.D. Poyser, 1999.

— INDEX —

Index

Index

Index

Index

Index

Index

— ABOUT THE AUTHOR —

Derek Gow is a farmer, nature conservationist and the author of *Bringing Back the Beaver* and *Birds, Beasts and Bedlam*. Born in Dundee, he left school when he was seventeen and worked in agriculture for five years. Inspired by the writing of Gerald Durrell, he jumped at the chance to manage a European wildlife park in central Scotland in the late 1990s before moving on to develop two nature centres in England. He now lives with his children, Maysie and Kyle, on a three-hundred-acre farm on the Devon–Cornwall border, which he is in the process of rewilding. Derek has played a significant role in the reintroduction of the Eurasian beaver, the water vole and the white stork in England. He is currently working on a reintroduction project for the wildcat.